OKANAGAN GEOLOGY

BRITISH COLUMBIA

MURRAY A. ROED AND JOHN D. GREENOUGH, EDITORS

Contributors

James Baker, Dave Capstick, Don A. Dobson,
George Ewonus, Paul Ewonus, Leanne Greenough, Rita C. Harris,
Brian Hughes, H.A. (Herb) Luttmerding, Ian R. Walker,
Norman Williams, Uthaya Uthayakumar

Third Edition 2014

First Printed 1995

Inquiries should be sent to:
Okanagan Geology Committee
c/o Earth & Environmental Science and Physical Geography
University of British Columbia Okanagan
3333 University Way
Kelowna, British Columbia V1V 1V7
Library and Archives Canada Cataloguing in Publication
Okanagan Geology : British Columbia / Murray A. Roed and John D. Greenough, editors ; contributors, James Baker, Dave Capstick, Don A. Dobson, George Ewonus, Paul Ewonus, Leanne Greenough, Rita C. Harris, Brian Hughes, H.A. (Herb) Luttmerding, Ian R. Walker, Norman Williams, Uthaya Uthayakumar. -- Third edition.
Includes bibliographical references and index.

ISBN 978-0-9699795-4-8 (pbk.)
1. Geology--British Columbia--Okanagan Valley (Region).
2. Geology--British Columbia--Kelowna Region. I. Baker, James, author
II. Greenough, John David, editor III. Roed, M. A., editor
QE187.R642014 557.11'5 C2014-906366-0
Front cover painting: M.A. Roed
Hardy Falls, Peachland B.C.
Back cover photograph: M.A. Roed
West Kettle Columns, Okanagan Falls Forestry Road
Graphics and Sketches: Hanny Muggeridge, M.A. Roed

Okanagan Geology Committee
Don A. Dobson, John D. Greenough, Brian Hughes,
Murray A. Roed, Norman Williams.

Distributed by Sandhill Book Marketing Ltd.
Unit 4, 3308 Appaloosa Way
Mill Creek Industrial Park
Kelowna, BC V1V 2W5
Tel: (250) 491-1446
Fax: (250) 491-4066
E-mail: info@sandhillbooks.com

Printed in Canada by Friesens

Cost: $24.95

SPONSORS

Printing of Okanagan Geology, 2014,
is sponsored by an Award from the Jérôme H. Remick III
Endowment Trust Fund, courtesy of the Canadian Geological Foundation

Kelowna Geology Committee Award Fund

MEMORIAM

This 3rd Edition of Okanagan Geology is dedicated to the memory of H.A. (Herb) Luttmerding and Dave Capstick. They were not only substantial contributors right from the start in 1995, they were our good friends.

PREFACE

This third edition of Okanagan Geology, British Columbia, is the result of sustained demand from the general public and their interest to learn and appreciate more about the land in which they live. The book is nearly the same as the 2004 edition but with some new photos, paintings and diagrams, plus it has been edited and revised somewhat due to new information. However, the highlight of the present book and a major reason for a third edition is an Addendum that presents the results of new geologic discoveries and major works in the Okanagan Valley in the last 10 years. The last time period when similarly significant works began was in the late 1950's when Hugh Nasmith mapped the surficial geology of the Okanagan (Nasmith, 1962), H. W. Little (1961) mapped the bedrock geology of part of this area, and A.G. Jones mapped the bedrock of the Vernon area (Jones, 1959). This is not to imply that other works have not been undertaken since, it just seems that the last ten years have been especially productive for the Okanagan. Subjects in the Addendum include the discovery of evidence for an early glaciation at least one million years old, a massive compilation of a bedrock geology map of the entire Okanagan Watershed has been done, unusual features on the bottom of Okanagan Lake were discovered in 2009, a new surficial geology map of the Kelowna area appeared in 2009, and the innovative engineering construction of the William R. Bennett Bridge that opened in 2008 is described by the engineer who designed it.

NOTE: Figures are numbered in order of appearance and are separated by a dash from the Chapter or other section (Appendix for example), Example, Figure 2-3 is the second figure in Chapter 3. Likewise Figure 5-B is the fifth figure in Appendix B, and Figure 3-AD is the third figure in the Addendum.

ACKNOWLEDGEMENTS

One of the major benefits of three geology books that have been published for the Okanagan is the Kelowna Geology Committee Award administered by our partner Okanagan College Foundation. All revenues from the books are endowed to this Award. It now totals approximately $80,000 even though we have distributed at least $20,000 to students at the College and UBCO since inception in 1995. The fund has increased by $50,000 in the last ten years and is now in the top 20 donor award list at the College. So, who do we thank for this? First we have to thank all of our scientific contributors who have volunteered their time to insure a modern, accurate and readable book. Secondly, we thank the Okanagan College Foundation for its administrative guidance and continued partnership and handling tax deduction issues for sponsors. Thirdly, and probably most importantly, we thank our public who purchased the book. Fourthly, we thank the many sponsors that have provided funds for the substantial expense of producing such a high quality book like Okanagan Geology. Tops amongst these is the Canadian Geological Foundation.

There are others to thank, talented people and agencies behind the scenes. Graphic Designers like Jack Thompson for the previous editions, and our present Graphic Designer John Wecels, and a special thanks to you John. Their input is critical for insuring the books are of high quality with innovative features that serve to make them that much more special. Then there is the distribution of the book and for that we salute Sandhill Book Distributors Ltd. of Kelowna under the expert direction of Nancy Wise. The book is available at no less than 70 outlets in the Okanagan and elsewhere. Good job Nancy! Last but not least is the media. We thank especially Judie Steeves, Capital News, for her continued interest and numerous articles, and Ron Seymour and Don Plant of the Kelowna Daily Courier. Mike Roberts of CHBC-TV is in a different category because of his incredible interest in all such geologic things and his insistence to go trekking around with us to every conceivable geologic landmark, with a patient camera man in tow. It has been fun Mike!

Finally, our sincere and special thanks to Debra Roed for her photography, assistance in the field, extensive editing of this revised edition, a lot of typing, and her constant encouragement right from the start.

TABLE OF CONTENTS

OKANAGAN GEOLOGY BRITISH COLUMBIA

CHAPTER NINE
GEOLOGIC LANDMARKS OF THE KELOWNA AREA

CHAPTER TEN

GEOLOGIC HAZARDS IN THE OKANAGAN VALLEY

CHAPTER ELEVEN

GEOTECHNICAL CONDITIONS IN THE KELOWNA AREA

CHAPTER TWELVE

MINERAL RESOURCES

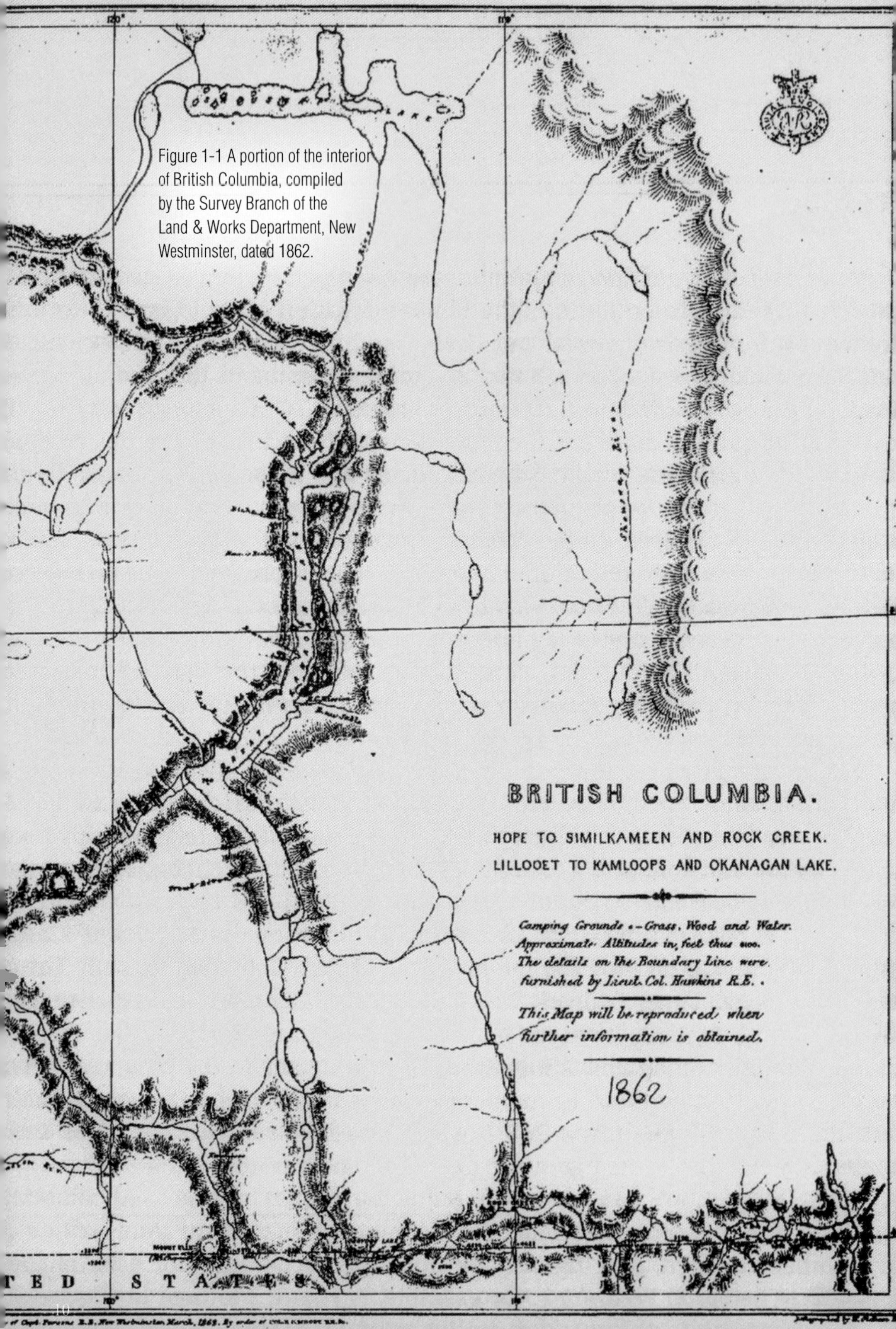

Figure 1-1 A portion of the interior of British Columbia, compiled by the Survey Branch of the Land & Works Department, New Westminster, dated 1862.

CHAPTER 1

HISTORY OF GEOLOGICAL STUDIES IN THE OKANAGAN VALLEY

by Murray A. Roed

The first geologist to visit the Okanagan Valley was George Mercer Dawson in 1877. The scope of his observations while travelling by foot, canoe and horseback was phenomenal. He commented on or described in detail every major geologic feature visible in the valley. Dawson was a Nova Scotia born scientist educated at McGill University. He joined the Geological Survey of Canada in 1872 and was first assigned to a joint British and American survey of the International Boundary. In 1875 he obtained an appointment to undertake a geological survey of British Columbia which led to a life-long devotion. Dawson arrived at Kelowna on July 3, 1877, and the next day visited the placer mines at Mission Creek (Cole and Lockner, 1989, p. 342). The creek below Gallagher's Canyon had been mined for gold for at least sixteen years. Dawson reported that coarse gold of two to three ounces per day per man was still being recovered there at that time, mainly from McDougall's claim (Dawson, 1879, p. 157b).

George Dawson, born in 1849, was the son of Sir William Dawson, a noted geologist and principal of McGill University. Dawson has been called the "little giant" of Canadian geology. His achievements are even more remarkable given his physical condition. He was no bigger than a boy of 12, had weak lungs, and a bent and hunched back, all effects of a childhood illness. Yet in 1887, for example, he traversed over 164,000 square

Figure 2-1: Dr. George Mercer Dawson,
Courtesy of the Geological Survey of Canada (photo number GSC 311).

kilometres of northern British Columbia, and then explored the Yukon nearly a decade before the Klondike gold rush of 1896. He was an avid collector of plants and flowers, and described all of those found along his wide ranging routes. He discovered numerous fossil localities, and had a particular interest in the language and culture of North American Indians. In 1895 Dawson took over the Director's chair of the Geological Survey of Canada. He suddenly died in 1901.

Geological investigation in the Kelowna area appears to have been largely absent during the next fifty years. Various geologists with the British Columbia Department of Mines periodically examined scattered mine workings, but no formal studies were conducted. Exceptions to this include a study by L. Reinecke (1915) on the ore deposits at Beaverdell and a treatise that focused attention on the physiographic or geomorphic aspect of the Okanagan Highland. C.E. Cairnes of the Geological Survey of Canada began a study of mineral resources of the northern Okanagan Valley in 1930. He then systematically began mapping the Kettle River map sheet which today is part of the Penticton map sheet number 82E of the National Topographic System where Kelowna is situated. Cairnes emphasized the mineral deposits of the region and also produced the first bedrock geologic map that included the Kelowna area (Cairnes, 1937). The early thirties witnessed considerable excitement about an alleged oil discovery near the mouth of Mission Creek at Kelowna. A summary is given by a skeptical Cairnes.

Cairnes' map remained the only geologic map of the area until H.G. Little spent five months in 1958 and 1959 revising Cairnes' work. Little's work was probably initiated as a result of geologic mapping in the Vernon area by H.M.A. Rice in 1945 and 1946 which was continued by A.G. Jones from 1947 to 1951 (Jones, 1959). These studies were greatly facilitated by access provided by the Hope-Princeton Highway opened in 1944. Railroad access via the Kettle Valley Railway had been continuous since 1916.

Other studies in the Okanagan between 1930 and 1950 focused on the glacial deposits of the region. Inspired by Dawson's original descriptions of ice-rafted deposits and his final and correct conclusion that the area was affected by extensive ice sheets, numerous scientists began to decipher the surficial deposits and origin of the present day landforms. The "white silts" described by Dawson are exposed in cliffs along Okanagan Lake and have long attracted the curiosity of geological scientists (Flint, 1935; Mathews, 1944), and still continue to do so (Shaw and Archer, 1979).

Despite widespread interest and importance of surficial geology, or glacial geology, it was as late as 1962 before Hugh W. Nasmith published maps showing the distribution of these deposits in the southern Okanagan Valley (Nasmith, 1962). Nasmith's maps remain widely used even though they are reconnaissance in nature with a scale of 1:126,720. Agricultural soil scientists have since mapped the soils of the Kelowna area in considerable detail (e.g. U. Wittneben, 1986), but little has been added to the glacial geology inventory since Nasmith. Exceptions to this include Stuart S. Holland's synthesis of British Columbia's landforms which is still the main authority on the physiographic divisions of the Okanagan Valley area (Holland, 1964) and the surficial geology map of Fulton (1965).

Beginning about 1960, an explosion of geological work began in British Columbia that in most cases relates only in a general way to the Okanagan. Groundwater resources were inventoried for the Okanagan Basin Study during the early 1970's. This included the first seismic profile across the valley north of Vernon. Soils maps of most agricultural areas were produced during the 1960's. Maps of the terrain and soils for the Canada Land Inventory project were completed during the 1970's. Recent specific and/or comprehensive studies for the Kelowna area are restricted to bedrock geology open file maps by Templemann-Kluit (1989), and an inventory of mineral occurrences by Meyers and Taylor (1989).

Volcanic rocks of Tertiary age in the Kelowna area have been of particular interest to Neil Church of the British Columbia Geological Survey (Church, 1980 and 1981). Church's maps are the only detailed bedrock geology maps ever published for the Kelowna area. W.H. Mathews of the University of British Columbia reported on plateau volcanics of late Tertiary age in uplands east of Kelowna (Mathews, 1988).

Other contemporary geologists who have made major recent contributions to the geology of the Okanagan Valley are Wheeler of the Geological Survey Canada, editor of a new bedrock geology map of the Cordillera, Okulitich for regional bedrock mapping in the North Okanagan, Parish and colleagues for their work on age dating and structural evolution of the Shuswap Terrane, Monger for his work on plate tectonics, Christopher for work on uranium and Beaverdell silver, Fulton (1975) for key discoveries related to the early glacial history of the area, and Eyles et al (1990) for an illuminating look at the sediment fill of Okanagan Lake.

From 1980 to 2003 a multidisciplinary team of geologists and geophysists conducted deep seismic reflection probes of the crust below the Cordilllera and other parts of Canada. This study is called Lithoprobe, and findings from this project are relevant to the Okanagan.

A summary of further work is provided in this revised edition, and is given in the Addendum at the end of this book.

EARLY HISTORIC MAPPING IN THE OKANAGAN

by Rita C. Harris

The map shown in Figure 1-1 of this book is by no means the earliest of the interior of British Columbia. The map represents a compilation of work done up to1862. It was drawn by J. Conroy, lithographed by W. Oldham, under the direction of Captain Parsons by order of Colonel R.C Moody, all of who belonged to the Survey Branch of the Lands and Works Department and were officers of the Royal Engineers. The assistance of Lieutenant Colonel Hawkins is also acknowledged.

One of the earliest maps of the interior of British Columbia was produced in 1827. It was drawn by the Chief Trader for the Hudson's Bay Company, Archibald McDonald of Fort Langley and Kamloops, and shows various territiories of first nations trading partners. The title was "A Sketch of Thompson's River District, 1827". The next map of consequence was again of Thompson's River District, this time by CT Samuel Black of Kamloops. Drawn at a scale of 1: 1,000,000, it compiled information from several years of travel, and dates approximately to 1835. It also records the passage of botanist David Douglas through the area and shows a number of geographic names on the west side of Lake Okanagan reflecting the route of the Brigade Trail (Harris et al, 1989). In May of 1867, A.C Anderson completed a manuscript of his travels between 1832 and 1851 with a map drawn at a scale of one inch to ten miles. This was one of the later maps of the colonial period in British Columbia but it was never published.

This "background" map, published in 1871, represents the classic 19th century map of the interior of British Columbia. It was drawn by J.B. Launders, a veteran of the Lands and Works Department. He arrived with Colonel Moody's Royal Engineers in 1858. J.W. Trutch was Launders' supervisor, Chief Commissioner of Lands & Works, equivalent to a Minister in modern Cabinets (see Figure 11-4).

CHAPTER 2

GEOLOGIC TIME

by Murray A. Roed

Geologists estimate that the earth formed about 4.6 billion years ago. The oldest rocks preserved on Earth are about 4.0 billion years old. They mark the beginning of the Precambrian which extends to 570 million years ago when fossilized animal life becomes abundant. Most of the rock material comprising the continents formed at this time. Rocks as old as 2.0 billion years occur on the eastern side of the Okanagan Valley though those at Kelowna are probably younger.

GEOLOGIC TIME INTERVALS

ERA	PERIOD	EPOCH	Millions of Years Ago
CENOZOIC	QUATERNARY	Holocene	(Recent) 0.01
		Pleistocene	2.0
	TERTIARY	Pliocene	5.0
		Miocene	24
		Oligocene	37
		Eocene	58
		Paleocene	66
MESOZOIC	CRETACEOUS		144
	JURASSIC		208
	TRIASSIC		245
PALEOZOIC	PERMIAN		286
	PENNSYLVANIAN		320
	MISSISSIPPIAN		360
	DEVONIAN		405
	SILURIAN		438
	ORDOVICIAN		505
	CAMBRIAN		570
PRECAMBRIAN			birth of planet 4600

Figure 1-2: Geologic time divisions

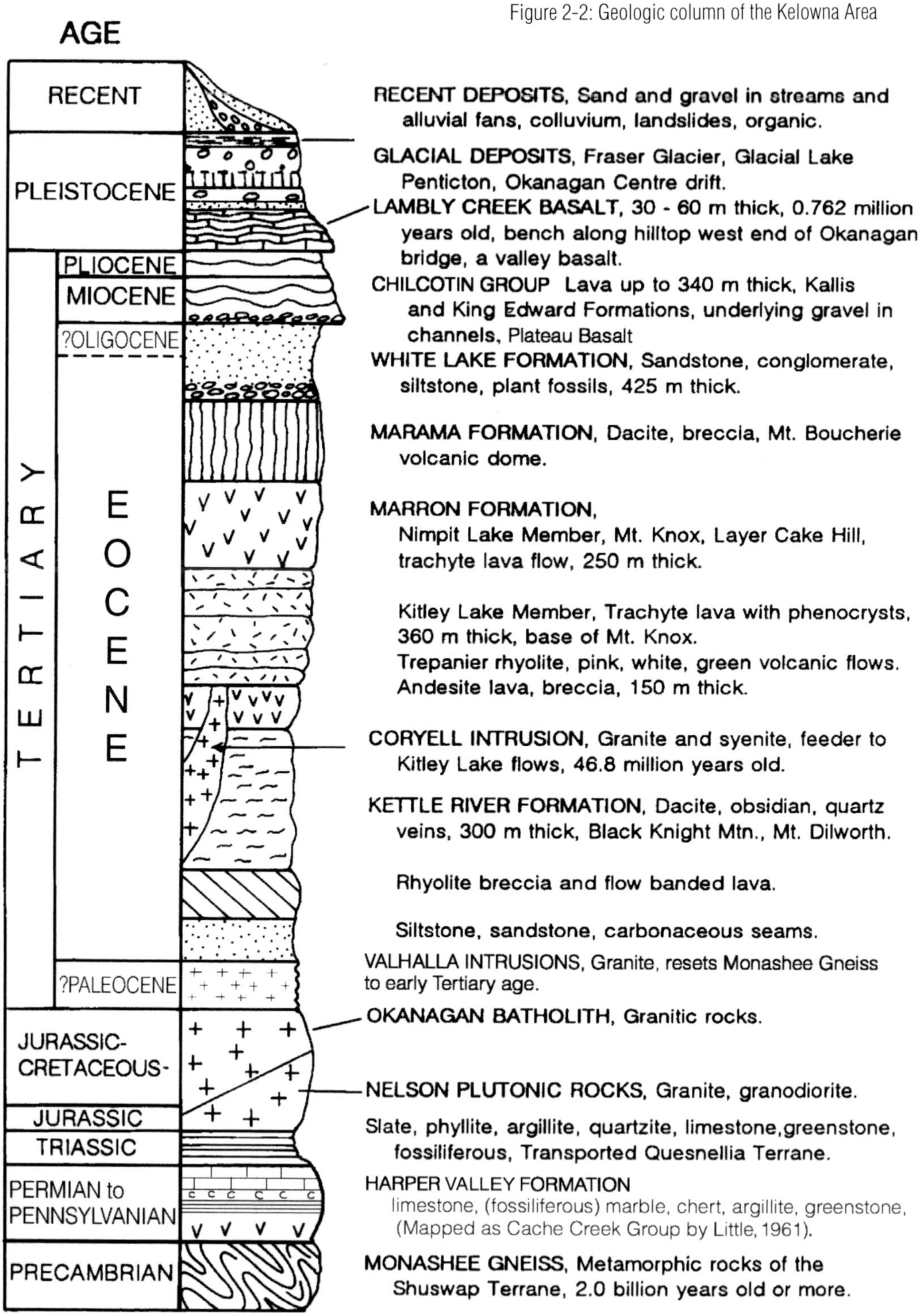

Figure 2-2: Geologic column of the Kelowna Area

The last 570 million years of geologic time forms the Phanerozoic Eon. It is formally divided into three major Eras, from oldest to youngest the Paleozoic, Mesozoic and Cenozoic (Figure 1-2). These Eras are divided into Periods (e.g. Tertiary) and the Periods into Epochs (e.g. Eocene). During the Paleozoic Era, from 570 to 245 million years ago, a great diversity of marine life and environments waxed and waned across the Earth. Animals and plants adapted to life on land and at the end a supercontinent, Pangea, had formed.

The Mesozoic Era spans from 245 to 66 million years ago and represents a time of major change in life forms on Earth. For example, dinosaurs flourished in swamplands of Alberta and Saskatchewan. Pangea began to break apart forming today's continents and causing mountains to start forming in western North America. The Okanagan however, was the site of mountain building with rocks being folded, broken, uplifted, transported, heated and injected with molten rock from deep in the Earth.

The Cenozoic Era spans from 66 million years ago to the present. During this time, which includes the locally important Tertiary Period, many of the visible geologic features in the Okanagan Valley were formed. The Ice Age (the Pleistocene Epoch) occurred near the end of the Cenozoic, and we live in the Holocene or Recent Epoch of the Cenozoic Era.

GEOLOGIC COLUMN OF THE KELOWNA AREA

The main geologic deposits, and their respective ages, that are found in the Okanagan, typified by the Kelowna area, are shown as a geologic column in Figure 2-2 compiled from Templemann-Kluit (1989), Little (1961), and Church (1980). From oldest to youngest these are:

PRECAMBRIAN

Okulitch suggested using the name Shuswap Metamorphic Complex for a Precambrian unit containing three parts. The Monashee Complex is the older part lying north of Vernon and east of the Okanagan Valley. The second part is referred to as the Shuswap Complex and lies largely north of Vernon and west of the Okanagan Valley. The third part is called the Okanagan Plutonic and Metamorphic Complex and extends south of Vernon on the eastern side of the Okanagan Valley and presumably would include Okanagan Mountain Park. Added to the confusion is that Templeton-Kluit has called the metamorphic rocks in the park Okanagan Gneiss. In this book only the general term Monashee Gneiss is employed.

Parts of the Monashee Gneiss were originally formed 2.2 to 2.3 billion years ago and represent the oldest rocks in British Columbia (Armstrong et al., 1991). Monashee Gneiss in the Kelowna area, referred to as the Okanagan Gneiss by Templeton-Kluit (1989), is probably younger but specific dates are not available. Monashee gneisses form a belt of highlands in central British Columbia extending from the Canada/United States border to Valemount, near Jasper National Park. They once formed the Pacific edge of an old continental nucleus of North America known today as the Precambrian Shield. For much of their existence these rocks were deeply buried (10-40 kilometres) and subjected several times to great heat and pressure during mountain building events. As a result the original chemical elements of the rocks have been repeatedly rearranged to form new minerals producing the banded metamorphic rock known as gneiss. How these deeply buried rocks came to occupy a substantial surface area around Kelowna is of profound interest, as discussed in Chapter Four.

PALEOZOIC

Recent investigations (Daughtry et al., 2000) have confirmed the presence of Late Paleozoic (Pennsylvanian to Permian) rocks of the Harper Valley Formation in the northwest corner of the Kelowna map area, also mapped by Little (1961). Rock types range from greenstone to slate, siliceous (hard) argillite, and distinctive white, coarsely-crystalline, massive limestone (Read and Okulitch, 1977). The limestones appear to outcrop along Westside Road. Small, rounded marble hills near Okanagan Landing at Vernon contain productellid fossils and may be similar. For a discussion of the hotly-contested, plate tectonic significance of the sedimentary rocks, see Chapter Four.

MESOZOIC

It is unclear whether there are Mesozoic Era sedimentary rocks in the Kelowna (this book) map area but outcrops of Triassic Period marine siltstone and shale occur 35 km to the north at Salmon River. There are, however, granitic intrusive rocks of Jurassic to Cretaceous age at Kelowna and these have been extremely important because they led to ore deposits such as the Brenda Mine.

CENOZOIC

Many geologic features in the Okanagan area formed during the Cenozoic Era which started 66 million years ago. During the first

period of the Cenozoic, the Tertiary Period, Kelowna was often volcanically active. Most local surficial deposits formed during the Quaternary Period, which begins with thc Pleistocene Epoch (Ice Age), and ends with the Holocene Epoch (the last 10,000 years).

Tertiary

Rocks of Tertiary age are widespread in the Kelowna area and are primarily volcanic in origin. The rocks make up the Kettle River, Marron and Marama Formations which belong to the Penticton Group (Church, 1980 and 1981). Well known landmarks such as Dilworth Mountain, Knox Mountain, Layer Cake Hill, Mount Boucherie, and Black Knight Mountain formed from local volcanic activity fifty million years ago. As volcanism waned river delta and stream deposits of the White Lake Formation partially buried the previous volcanic landscape.

The youngest Tertiary deposits are basalt lava flows of the Chilcotin Group which poured out across uplands and flooded ancient river valleys. Today the lava flows, commonly with underlying river gravels, form topographic high points due to their resistance to erosion. Radiometric dating suggests at least two episodes of volcanic activity (Mathews, 1988). Flows belonging to the King Edward Formation erupted eleven to twenty million years ago and occur on the upland north of Mission Creek toward Vernon. Kallis Formation flows erupted less than seven million years ago and can be found in the Big White Mountain and Hydraulic Lake areas.

Figure 3-2: McIntyre Bluff and Vaseux Lake, acrylic painting by M. Roed

Quaternary

Various deposits and landforms were produced during several glaciations and interglacial events of the Pleistocene Epoch, the first part of the Quaternary Period beginning about two million years ago. Minor volcanic activity is also known from this time interval. The Lambly Creek basalt on the west side of Okanagan Lake erupted 762,000 years ago. Also, some plateau basalts in the Hydraulic Lakes area east of Kelowna poured out of fissures one to three million years ago.

At the beginning of the Holocene Epoch, the last part of the Quaternary, 10,000 years ago, a huge glacial lake occupied the valley. When it drained sediments that had accumulated at its edge were exposed and eroded forming "benches" or "silt bluffs" characteristic of the Kelowna and Penticton landscape.

MCINTYRE BLUFF, (Figure 3-2) is situated on the western side of Okanagan Valley, halfway between Oliver and Okanagan Falls. Vaseux Lake, part of the Okanagan River system, abutts this prominent landmark. The bluff exposes a 500 metre high wall of granite gneiss, shot through with thin layers of quartz and other mineral veins. The rock is known as the Vaseux Formation and belongs to the Shuswap Metamorphic Complex of Precambrian/ Devonian age. This cliff is the steepest and most dramatic in the Okanagan Valley, and the valley itself is narrowest and most constricted at this spot.

Other geological dramas have been identified at this site. The rocks have been subject to major tectonic (mountain building) events that served to fold the rocks into great overturned fold structures at least 10 kilometres wide. The main structure is called the McIntyre Bluff Fold. It is a huge recumbent to isoclinal syncline with a northeasterly trending fold axis. It is also referred to as a nappe. Although three periods of mountain building and folding are recognized in the South Okanagan by geologists John V. Ross and James S. Christie of the University of British Columbia, the main fold structure dates to at least Permian time, over 270 million years ago.

How was this landform carved out of this mountainous terrain? According to Dr. Hugh Nasmith, who studied the glacial history of the valley in the early 1960's, deep intensive erosion of the Okanagan was first accomplished by one or more rapidly flowing valley glaciers. It is very likely that several periods of valley glacier erosion operated to repeatedly gouge out a path down the valley during the last one or two million years. It is known from other parts of the valley that the bedrock floor has been eroded so deeply that today it is far below sea level at an elevation of -640 metres (-2100 feet), 1000 metres below Okanagan Lake, for example. Overall, in fact, the Okanagan Valley has witnessed the removal of nearly three kilometres of its rock by glaciers and ancient rivers, which is almost twice as deep as that eroded from Arizona's Grand Canyon!

CHAPTER 3

PHYSIOGRAPHIC DIVISIONS

by Murray A. Roed

The landscapes of southern British Columbia are defined by major physiographic divisions which include the Coast Mountains, Thompson Plateau, Okanagan Highland, Shuswap Highland, Monashee Mountains, Selkirk Mountains, and far to the east the Purcell Mountains (Figure 1-3). Striking linear lake-filled valleys are common to most of these regions, an impressive example being the Okanagan Valley. The Okanagan Valley at Kelowna transects an eastern portion of an upland called the Thompson Plateau and part of the Okanagan Highlands. The region occupies the southern portion of the Interior System, specifically the Interior Plateau portion of the Southern Plateau and Mountain area. The origin of these physiographic units is broadly described below.

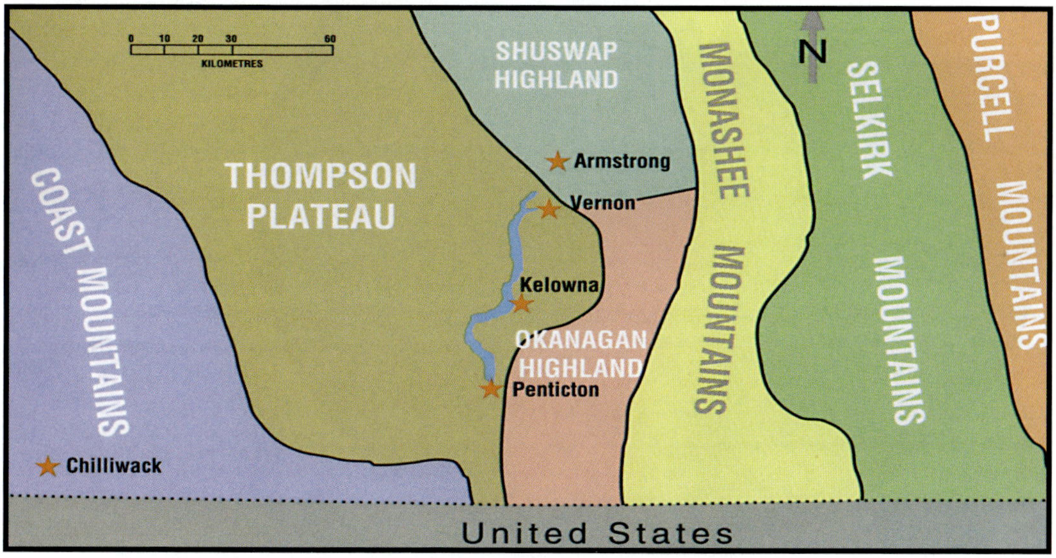

Figure 1-3: Physiographic Divisions of Southern British Columbia.

THOMPSON PLATEAU

Volcanic and sedimentary rocks underlying most of the Thompson Plateau at Kelowna and represent exotic terrane (they originated somewhere in the Pacific) that was added to the side of the North American continent as a result of North American and Pacific plate convergence during the Mesozoic (see Chapter Four). Later this terrane was pervasively intruded by magmas now forming granitic rocks and disturbed by Eocene volcanic activity. The bedrock making up the Thompson Plateau began to take its present physiographic or topographic form in late Eocene and early Oligocene times. Much of the interior of British Columbia was then uplifted in response to the gradual rise of the Rocky Mountains to the east and subduction of the Pacific Plate at the Continental Margin to the west. The plateau's broad rolling to gently sloping high level "plains" today resulted from extensive erosion of this uplift and represent an ancient "peneplained" surface. Remnants of these upland surfaces are now bordered by prominent valleys over 1000 metres deep. The valleys formed by gradual stream dissection of the ancient peneplain in the late Tertiary and early Quaternary Periods.

Near the end of the prolonged period of Thompson Plateau uplift and erosion, volcanic activity began again in the interior of British Columbia. Volcanism commenced as early as twenty million years ago and peaked ten to fifteen million years ago. Highly fluid basaltic lavas flowed out over many parts of the vast pre-existing peneplain topography forming "plateau basalts" referred to as the Chilcotin Volcanics. The lavas spread out in thin sheets, travelling for great distances along flat-bottomed valleys of the ancient peneplain drainage system. These basalt flows tend to exaggerate the present day plateau surface of the Thompson Plateau in places.

Figure 2-3: Typical Escarpment of the Thompson Plateau near Vernon & Oyama

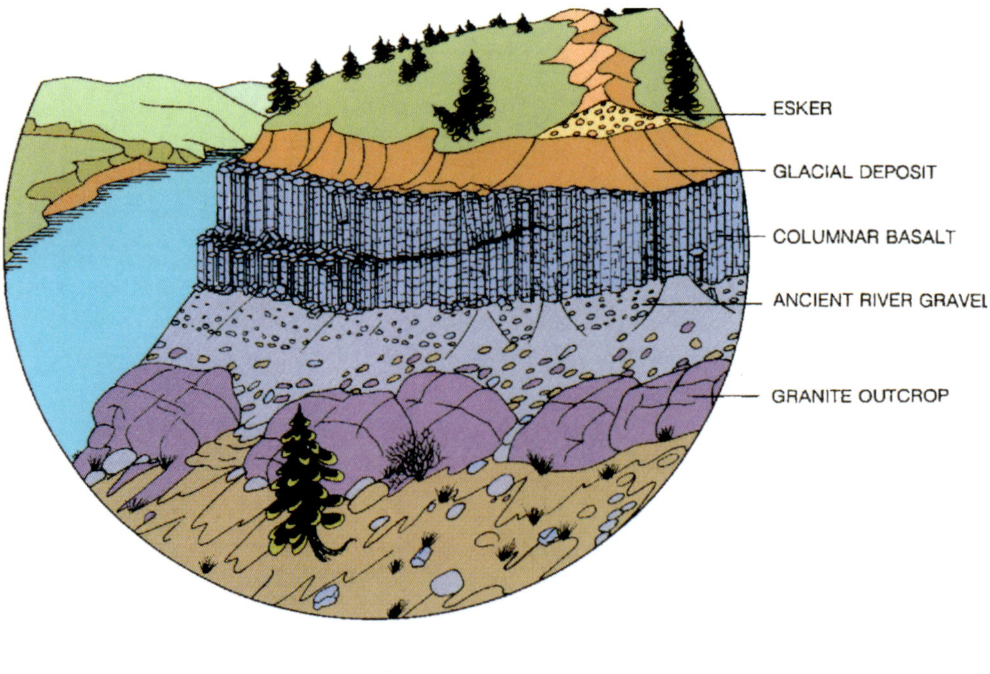

The Chilcotin lava obliterated a regional river system that had developed on the peneplain of central British Columbia. Details of the drainage pattern are not known but geologists do know that a province-wide drainage system existed that was far different from what exists today. Even the mighty Fraser River drained northward during this time. Remnants of this river system exist today as gravel deposits preserved below plateau basalts. The gravel deposits contain uranium and placer gold in places (see Chapter 12). A geologic profile of the plateau basalts and river gravels typically exposed along valley walls cutting through the Thompson Plateau is illustrated in Figure 2-3.

OKANAGAN HIGHLAND

Rolling to somewhat rugged mountains of the Okanagan Highland flank the Thompson Plateau to the east (Figure 1-3). The area is dominated by metamorphic rocks of the Monashee Gneiss and granitic rocks. During the Eocene and Oligocene Epochs the rocks were thrust up along a great fault located partly along Okanagan Lake at Kelowna. The Okanagan Highland represents the roots of an ancient mountain system that emerged during the Tertiary Period (Chapter Four).

The granitic rocks include Jurassic to Cretaceous granite and diorite igneous rocks referred to as the Okanagan Batholith and Middle Jurassic Nelson Plutonic Rocks. These rocks intrude the Monashee rocks south of Beaverdell and in Okanagan Mountain Park. Heat from another substantial group of granitic rocks of Tertiary age referred to as the Valhalla Intrusives has also affected this area and have reset the "radiometric clock" of the Monashee Gneiss to reflect an early Tertiary period of metamorphism (Mathews, 1988). Other Okanagan Highland rocks include minor patches of Upper Paleozoic to Lower Mesozoic Pacific ocean volcanic and sedimentary rocks of the Nicola Group that were "sutured" to and thrust onto the North American continent during the Mesozoic.

Like the Thompson Plateau, the Okanagan Highland is locally capped by late Tertiary, Chilcotin Group plateau basalts. At some locations sand and gravel deposits up to 50 metres thick and 100 to 300 metres wide underlie the basalts. The sedimentology, distribution, orientation, and elevation (1000 metres above the present Okanagan Valley bottom) of these gravel deposits give evidence for a high energy drainage system bearing no relationship to the present or earliest Quaternary (pre-Ice Age) drainage patterns. Excellent examples of these ancient channel deposits can be viewed in the Grizzly Upland along King Edward Creek southeast of Vernon, east of Oyama, along West Kettle River on the way to Big White Mountain, and in some exposures along the Kettle Valley Railway above Kelowna.

EFFECTS OF GLACIATION ON PHYSIOGRAPHY

The Okanagan Highland experienced the same general uplift and erosion as the Thompson Plateau and both were heavily abraded by at least two and perhaps four Cordilleran glaciers during the Pleistocene (Chapter Five). Uplands today are mantled with a thin glacial moraine related to the melting and retreat of the last Cordilleran ice sheet, the Fraser Glacier.

Valleys in the Okanagan Highland and Thompson Plateau are dominated by interbedded moraine and outwash deposits in well formed terraces. Some valleys, notably

Mission Creek and Okanagan Lake valley, show extensive silt bluffs and linear terraces composed of silt, clay and sand related to glacial lake sedimentation.

Several spectacular rock-rimmed canyons with tumbling mountain streams mark the eastern part of the Thompson Plateau east of Kelowna. Examples include KLO Creek, Myra Canyon and Bellevue Creek. The Crawford Trail and portions of the Kettle Valley Railway offer excellent opportunities to view representative geologic features of the Okanagan Highland.

These canyons and many present valleys were formed approximately thirteen thousand years ago as a result of deep erosion from meltwater during deglaciation. Mission Creek and the upstream portion of Gallaghers Canyon (Figure 3-3) offer the nearest opportunities to view the effect of the extensive erosion that occurred.

Figure 3-3: A near vertical cliff eroded by Mission Creek since the end of glaciation is one of the many geologic features of the Scenic Canyon Regional Park and Gallagher's Canyon area in Kelowna. Acrylic painting by M. Roed.

Figure 4-3: One of the bridges along Mission Creek Greenway.

MISSION CREEK GREENWAY

by Murray A. Roed

The Mission Creek Greenway (Figure 5-3) will eventually be a 22.5 kilometre linear park along the creek from the shore of Okanagan Lake to Mission Creek Falls (Figure 11-9). The geologic features along the planned trail are numerous and diverse, many of which are displayed in this book. They include Layer Cake Hill and the Pinnacle, spectacular fold structures in the White Lake Formation (Figure 5-4), haunting hoodoos in glacial deposits and sandstone, exposures of ancient river systems (Figure 13-9 and 15-9), and gigantic boulders dropped by the Fraser Glacier. This has been an outstanding community project, and a monumental dedication to how much Okanagan citizens care for their environment. See pages 85 - 91 for more.

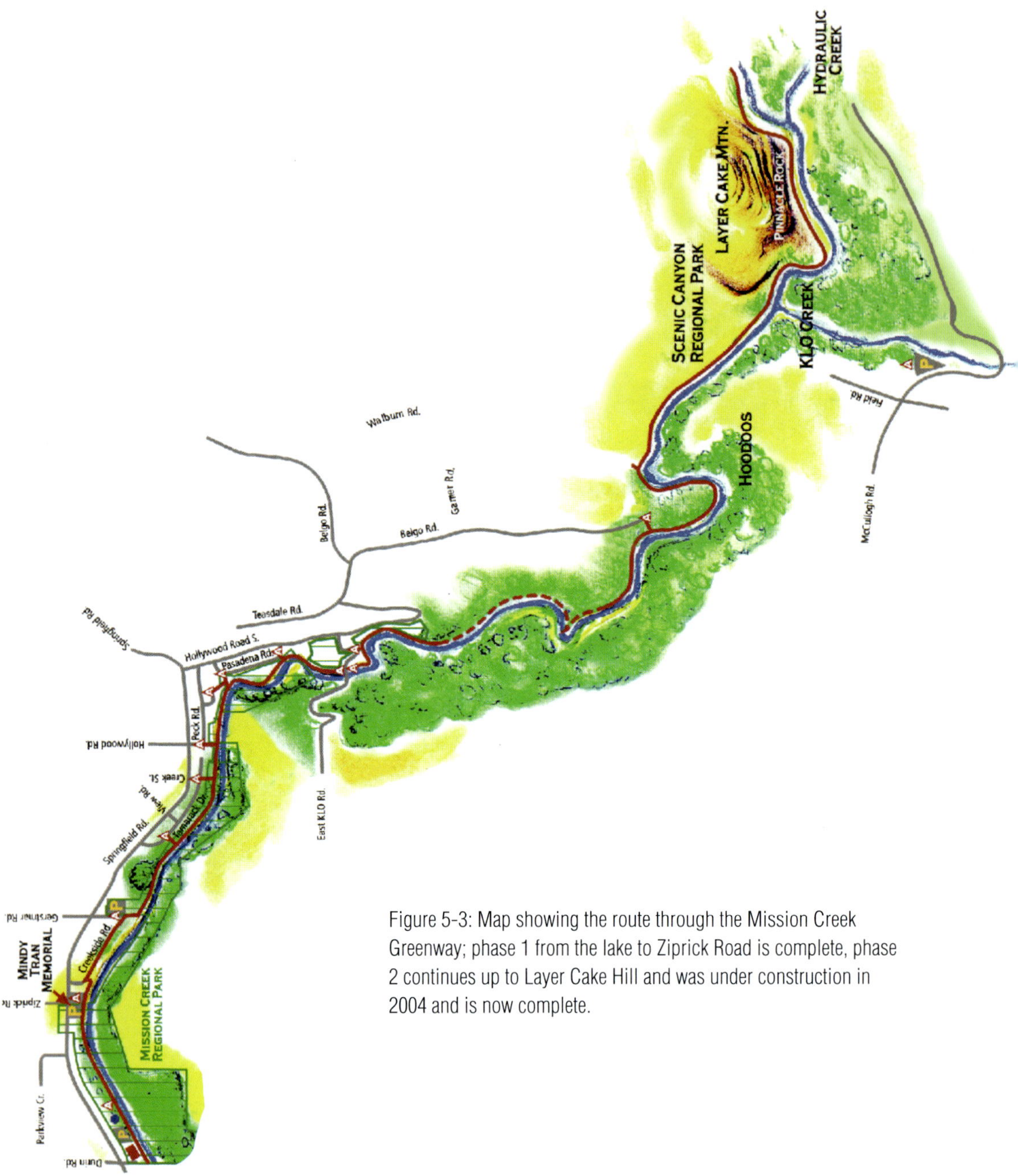

Figure 5-3: Map showing the route through the Mission Creek Greenway; phase 1 from the lake to Ziprick Road is complete, phase 2 continues up to Layer Cake Hill and was under construction in 2004 and is now complete.

CHAPTER 4

GEOLOGICAL HISTORY OF BEDROCK IN THE KELOWNA AREA

by John D. Greenough and Murray A. Roed

The principal geological events that formed the bedrock geology of the Okanagan Valley at Kelowna are summarized in this chapter. Discussion is restricted to pre-glacial events. A generalized bedrock geology map of the Kelowna area appears in Figure 1-4 and a new map is in the Addendum.

PLATE TECTONICS OF WESTERN CANADA

Over the last forty years the theory of plate tectonics has revolutionized geologic thought. According to this theory a rigid outer layer of the Earth, composed in part of the Earth's crust, is divided into plates. Like ice on a pond these plates "float" on a partially-molten or semi-solid portion of the mantle. The mantle moves (convects) and the overlying plates are carried along as if on a conveyor belt. In places, particularly at ridges in the ocean basins, plates move away from one another forming new basaltic ocean floor. Where plates collide, ocean floor may be pushed beneath other ocean floor or below continental crust. This process, known as subduction, results in volcanoes (e.g. Cascades and Aleutians), folding, faulting and metamorphism. In short, mountain building occurs. It is also possible for two continental plates to collide (e.g. Himilayas) and for plates to slide past one another (e.g. San Andreas fault).

Figure 2-4 shows the present configuration of plates in western Canada. British Columbia forms part of the North American plate. Most of the Pacific Ocean is underlain by the Pacific plate. New ocean floor is being produced just off British Columbia at the Juan de Fuca and Explorer ocean ridges. This creates two small plates which are being subducted below western Canada (Figure 2-4). They are young, still-warm and buoyant, and resist being subducted, and this leads to the periodic (every few hundred years) major earthquakes that people living on the coast live in fear of.

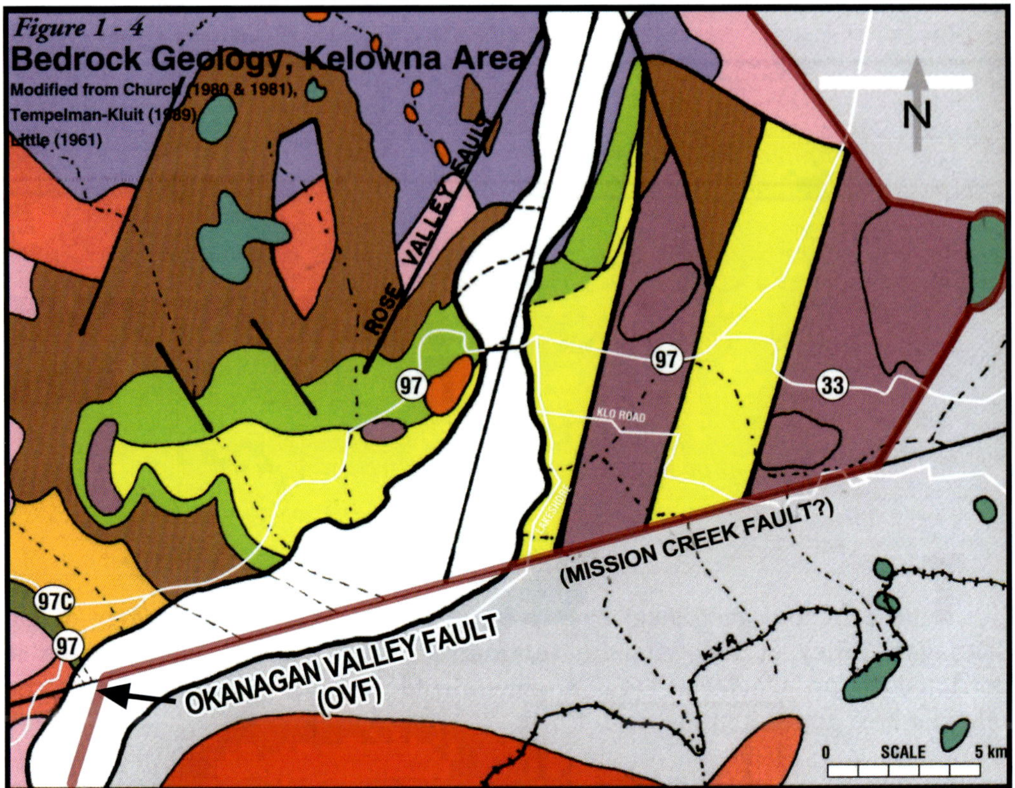

QUATERNARY
Lambly Creek Basalt
- dacite, vesicular basalt, a valley lava flow, Lakeview Heights

MIOCENE - PLIOCENE
Plateau Basalt, Chilcotin Group
Kallis Formation, King Edward Formation - andesite and basalt flows up to 20m thick - underlain by gravel

EOCENE
White Lake Formation
- carbonaceous siltstone, sandstone, conglomerate, volcanic conglomerate, thin to massive bedded

Penticton Group
Marama Formation
- grey intermediate volcanic flows, Layer Cake Hill, Mount Boucherie, Mount Dilworth.

Nimpit Lake Member, Maron Formation
reddish volcanic flows

Kitley Lake Formation
buff intermediate volcanic flows

Trepanier Rhyolite
- White, pink and green rhyolite volcanic flows

PALEOCENE
Valhalla Intrusives
Granodiorite
intrusions in Okanagan Mountain Park
- uncertain age

CRETACEOUS and/or JURASSIC
Okanagan Batholith
- light grey granite, grandiorite, intrusive rock

MIDDLE JURASSIC
Nelson Plutonic Rocks
- foliated granite and grandiorite, intrusive rock

TRIASSIC
Nicola Group
- slate, phyllite, argillite, greenstone; transported Quesnellia Terrane

PALEOZOIC
Pennsylvanian - Permian
- Harper Valley Formation, marble, argillite, chert, greenstone; fossiliferous

PRECAMBRIAN
Monashee Gneiss, Shuswap Terrane
- strongly foliated, grey granitic gneiss

Figure 1-4

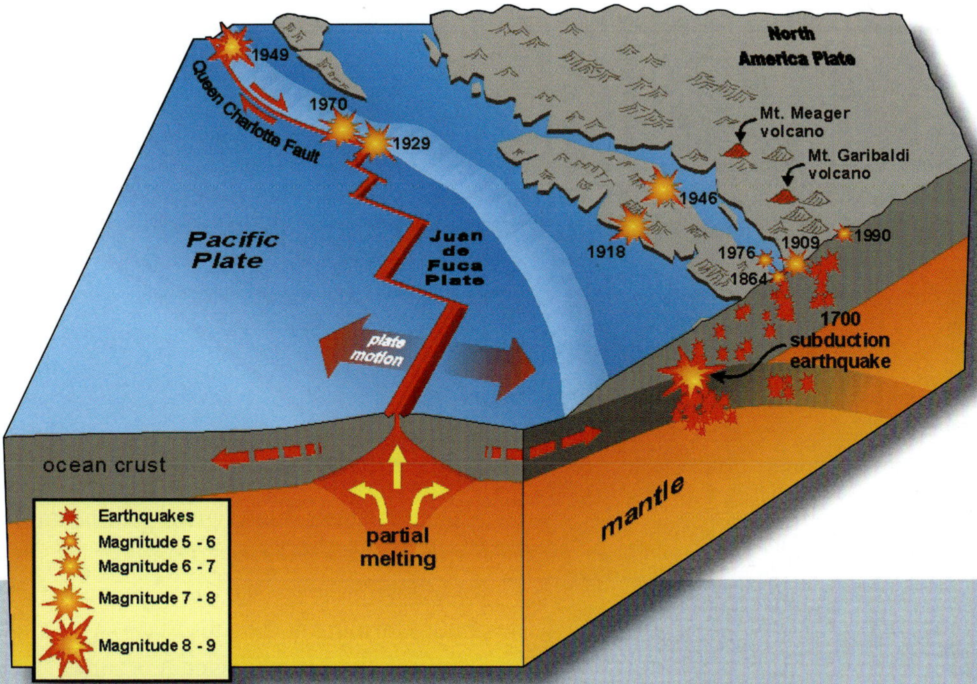

Figure 2-4: Simplified cross-section of southwestern British Columbia illustrates the Juan de Fuca and Explorer plates sliding beneath the North American plate off the west coast of Vancouver Island. Dates refer to major earthquakes.

PLATE TECTONICS AND THE OKANAGAN

Over the past 200 million years, the collision of the North American Plate and plates from the Pacific has resulted in subduction of ocean floor below North America (Figure 2-4). However, from time to time the subduction zones have become "clogged" and portions of plates have been successively "pasted" to the side of the continent. In some cases they were thrust up onto the continent and transported for hundreds of kilometres. The result is a series of northerly trending "terranes" of foreign or exotic origin. Some originated in the southwest Pacific whereas others have always been part of the North American continent. The accreted (added) or transported terranes may account for over half of the width of the Canadian Cordilleran mountain range and perhaps fifty percent of its volume. Canadian scientists (e.g. Clowes et al, 1995) working with the LITHOPROBE project have used seismic (sound) waves to study the internal structure of the mountain belt (Figure 3-4). Their work shows that most of the land west of the Okanagan Valley has been added to, or moved along, and/or transported over, the old (Precambrian Monashee gneiss) western edge of North America.

One of the transported terranes, Quesnellia, is represented in the Kelowna area. Conventional views of Quesnellia are that it represents a thick sedimentary and volcanic bedrock assemblage originally deposited in a volcanic island arc environment somewhere in the Pacific Ocean basin and accreted to North America during the Jurassic (Gabrielse et al., 1991). Recent work has challenged this

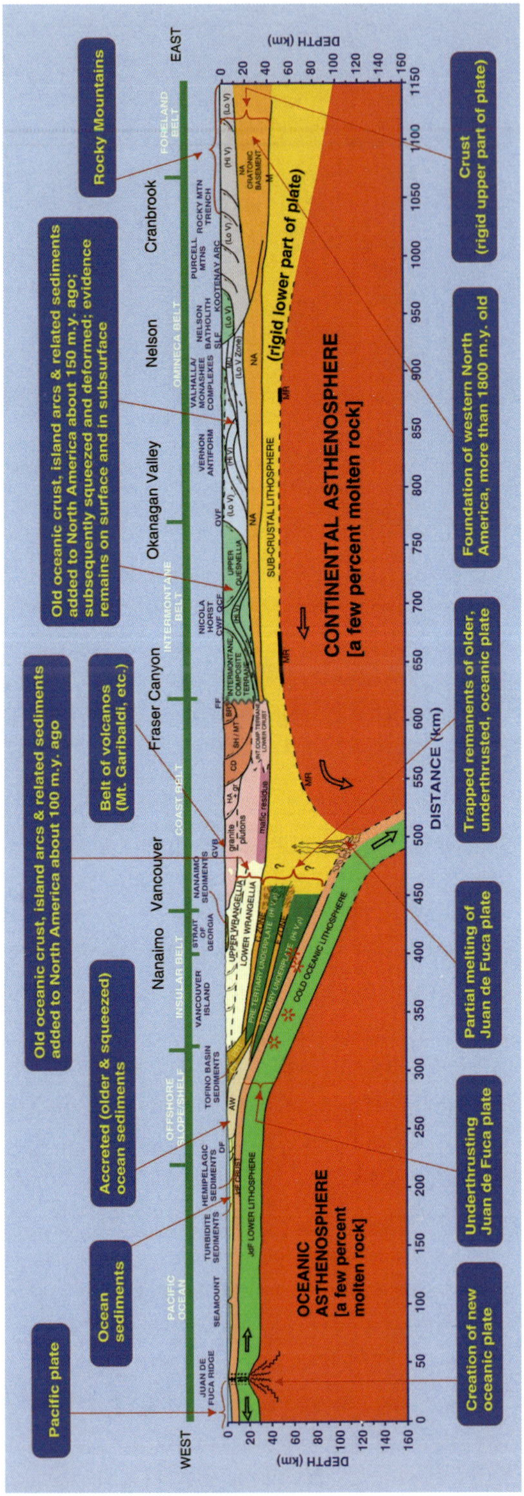

Figure 3-4: LITHOPROBE Cross Section of the Canadian Cordillera (Courtesy of Dr. Ron Clowes, Director, Lithoprobe, UBC)

hypothesis. For example, the Pennsylvanian to Permian sediments of the Kelowna area (Harper Valley Ranch Formation) may have been deposited on the North American continental shelf (Daughtry, 2000) and may not be substantially transported. Chemical studies of Triassic sediments near Vernon support derivation of the sediment from continental North America as opposed to an island arc environment (Unterschutz et al, 2002).

Subduction below the edge of continental North America has created episodes of volcanism at the surface with "granitic" plutons forming in the subsurface. As wet ocean floor is subducted it becomes heated causing water to be driven off. The buoyant, water-rich fluids rise and act like antifreeze causing melting in mantle rocks that underlie the continental crust. In turn, the basaltic magma rises and intrudes the crust causing it to melt. The heat added to the crust also results in the intruded country rocks becoming metamorphosed. Figure 3-4 shows that these processes occurred in the Kelowna area during the Middle Jurassic and Jurassic-Cretaceous Periods forming the Nelson plutonic rocks and the Okanagan batholith. Similar processes occurred again in the Tertiary (see below).

UPLIFT OF THE SHUSWAP TERRANE AT KELOWNA

Uplift of the Monashee Gneiss was a major event in the Okanagan and for British Columbia as a whole. This period of mountain building is referred to as the Columbian Orogeny.

As described previously Monashee gneisses of the Okanagan Highland represent the ancient edge of North America and were deeply buried below Quesnellia Terrane rocks during the Mesozoic. Perhaps during Cretaceous time the Highland began to rise. At the same time extension caused major portions of the Quesnellia Terrane to begin sliding off of the central rising mass. This westward sliding along the Okanagan Fault (see below), combined with erosion of Quesnellia Terrane rocks has served to bring Monashee gneisses of the Shuswap Terrane, once buried ten to forty kilometres, to the surface (Figure 4-4).

THE OKANAGAN VALLEY FAULT

The Okanagan Valley Fault is a major zone of displacement in southern British Columbia (Tempelman-Kluit and Parkinson, 1986). It has been determined by geophysical studies to be at least twenty kilometres deep. The fault is part of a system of north trending faults that can be traced for much of the length of the province. This discovery has relegated the fault to the status of one of British Columbia's largest Earth structures. Its influence on the geologic development of this part of the province has been profound.

The Okanagan Valley Fault played an important role in plate movements by acting as a major zone of rupture that "uncovered" the Monashee Gneiss Terrane. The fault can be traced for at least 200 kilometres and closely follows the Okanagan Valley. In places it diverges from the trend of Okanagan Lake, notably where it coincides with Kalamalka Lake, and at the intersection between Mission Creek valley and Okanagan Lake at Kelowna. Here it appears that the fault has been displaced by an east-west trending cross-fault mapped by Little (1961) and here called the Mission Creek fault (Figure 1-4). This second fault may be responsible for the huge bend in Lake Okanagan, the so-called "knee" of the lake. Some details of the deformation along this fault can be seen in Gallagher's Canyon (Figures 5-4 and 12-9). Other displacements are shown in Figure 11-10.

The Okanagan Valley Fault became most active in the Eocene Period (approximately sixty million years ago) and probably created a rugged landscape not unlike the Great Rift Valley of Africa. Initial extension and movement on the fault likely coincided with the inception of Eocene volcanic activity that produced calderas at Penticton, Summerland, Kelowna, Vernon and other locations in the interior of the province.

EARLY TERTIARY VOLCANICS

Active volcanoes and large rivers dominated the rift-valley landscape of the Okanagan during the Eocene Epoch, 50 to 60 million years ago. The rock record indicates that the volcanoes produced both lava flows and ash deposits not unlike the Cascade volcanoes Mount St. Helens and Mount Baker. Magma (molten rock) building these volcanoes had intermediate to high concentrations of silicon, an element that makes magma thick and sticky, and the volcano potentially explosive. A typical volcano went through one or more cone building stages during which both lava flows and ash deposits were repeatedly produced. Then, after a period of dormancy during which gases built up in the underlying magma

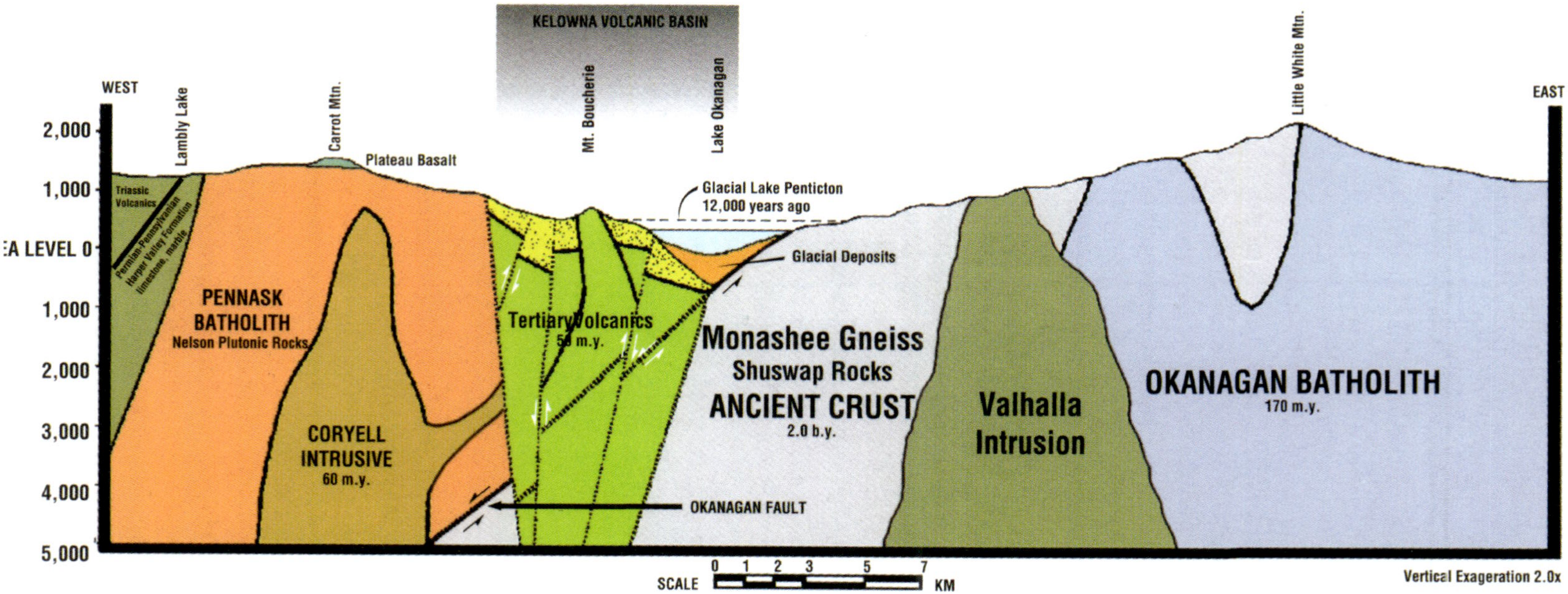

Figure 4-4: Cross section through Kelowna Basin showing the general distribution of major rock types and examples of the main faults. Dashed lines are faults or breaks along which movement (arrows show relative direction) has taken place in the past.

Shuswap Rocks are composed of the Monashee Gneiss. The Coryell and Valhalla Intrusives and the Okanagan and Pennask Batholiths are composed of granite, diorite and syenite. Volcanics in the faulted Kelowna Volcanic Basin include trachyte, andesite, basalt and others which belong to the Penticton Group. Included is the White Lake Formation (yellow) composed of siltstone, sandstone and conglomerate with coal layers and abundant plant fossils.

Figure 5-4: A northerly view across Mission Creek valley at Gallagher's Canyon reveals a series of overturned folds (dashed lines) in sandstone of the White Lake Formation. The volcanic rocks of Layer Cake Hill just to the east (right) appear to have been thrust westward causing these spectacular folds. These northerly trending folds have been cut by substantial movement along a major easterly trending fault, the Mission Creek fault (forming the "knee" of Okanagan Lake). The Mission Creek fault apparently offsets the Okanagan Valley Fault zone.

chamber, a catastrophic eruption blasted away much of the volcano leaving behind a large bowl-shaped depression known as a caldera. Subsequent volcanism and stream activity served to fill the caldera with more lava flows, ash deposits and volcanic-rich sediments. During the cone-building stage a hazard produced by the damp climate of the time was that snow built up on the peak of the volcano. Sudden melting during an eruption like the 1980 event at Mt. St. Helens yielded mudflows on the flanks referred to by geologists as lahars. Lahars cause extensive floods and widespread erosion, and may result in large volumes of sediment that choke drainage systems many kilometres downstream. Considering the volcanic threats, the propensity for flooding from abundant precipitation and the likelihood of earthquakes caused by movement on the Okanagan Valley Fault, Kelowna was a geologically hazardous spot during the Eocene!

AN EARLY TERTIARY RIVER

Gradually volcanism waned in the valley. The region was subjected to sedimentary infilling by a huge river system that deposited sediments of the White Lake Formation. Thick deposits of silt, sand, gravel and volcanic debris were deposited into the Kelowna basin partly burying the rugged volcanic terrain. Today this event can be envisioned at Boucherie Mountain where conglomerate and sandstone envelopes the base of the mountain (Figure 6-4).

This sedimentary event witnessed the emergence of forests that struggled to develop in the flood prone margins of the Kelowna basin. Evidence of this ancient forest can be seen in thin coal beds preserved in the White Lake Formation exposed along Highway 97 near Westbank (See Hardrock Trail, Appendix B, Stop 4, and number 21 in Figure 1-9). Fossil trees occur in an almost upright position, as if they were suddenly buried, and many sandstone beds contain well preserved plant fragments. These rocks outcrop in many places around Kelowna such as beside Mission Creek in Gallagher's Canyon, along Boucherie Road, in Glen Canyon Regional Park bordering Powers Creek, and along the western shore of Okanagan Lake south of the bridge (Figure 7-4).

Figure 6-4: An excellent exposure of the White Lake Formation occurs in Summerland along Highway 97 just below Sumac Winery. Here are massive volcanic conglomerates interbedded with thin bedded sandstone and lignite-rich layers. These different rock types record extremes in a river depositional environment ranging from rapidly deposited debris flows to placid overbank organic-rich floodplain deposition where lignite eventually formed.

Figure 7-4: Glen Canyon Regional Park displays many characteristics of the White Lake Formation. Along the picturesque Powers Creek here, sheer cliffs up to 100 metres high expose sandstone and conglomerate of the formation that have been eroded firstly by huge volumes of meltwater and then by the present creek which is still actively eroding its channel.

Following deposition of the White Lake Formation the Kelowna area underwent a prolonged period of regional uplift, folding and faulting. The Okanagan region was subjected to squeezing forces possibly related to the rise of the Rocky Mountains during Oligocene and early Miocene time from about 30 to 24 million years ago. Thus the older Eocene volcanic rocks and related sediments were folded and faulted. As explained above, good examples of the results of these mountain building forces can be viewed in Scenic Canyon Regional Park (Gallagher's Canyon) where sandstone of the White Lake Formation displays steeply upturned and even overturned beds (Figures 5-4 and 13-9).

This is when the Thompson Plateau and Okanagan Highland began to be eroded into a peneplain. The mountainous highland first became a major watershed that witnessed widespread erosion. One result was deposition of extensive deposits of sand and gravel that blanketed much of the foothills and adjacent plains of Alberta, 500 kilometres away. This activity ceased only when the Rockies began to rise in earnest by which time the Thompson Plateau and similar highlands in central British Columbia had been reduced to a broad peneplain with low relief.

A LATE TERTIARY RIVER SYSTEM

Because of a gradual northwestern continental drift, the Okanagan area as recently as 25 million years ago was 400 to 500 kilometres further south of its present latitude. The climate was distinctly tropical, with rainfall in the order of 700 to 1000 centimetres per year, compared to the present 55 centimetres for the Okanagan watershed. As mentioned in Chapter Two this caused development of a late Tertiary regional river system in central British Columbia. Numerous examples of this river system occur in the Kelowna area. The associated sand and gravel deposits have been mined for their placer gold content. Gold tends to be concentrated at the base of the deposits where streams eroded into the underlying bedrock. The sand and gravel deposits of the ancient drainage system also became hosts for uranium deposits (see Chapter Twelve). Uranium-bearing groundwater percolated through faults and fractures in the underlying bedrock and deposited its uranium when it encountered the near-surface chemical conditions of the river gravels. Subsequent basaltic lava flows of Miocene to Pliocene age (described below) have not only acted to preserve the river sediments, they form a protective cap on the uranium precipitated in the gravels. At least one potentially mineable uranium deposit is known in the area.

LATE TERTIARY PLATEAU VOLCANISM

A prolonged period of basaltic volcanism began in the interior of British Columbia approximately 20 million years ago and resulted in the formation of the Chilcotin Group plateau lavas that act as erosion-resistant caps on many hills in the Thompson Plateau and Okanagan Highland. Unlike the explosive Eocene volcanism where distinct volcanic edifices developed, the Chilcotin basalts passively emanated from long cracks or fissures that opened up in the crust. These highly fluid basaltic lavas spread out over the interior plateau and flowed down any river valleys or depressions existing at the time. The basalts remain relatively flat lying having been extruded after the cessation of tectonic forces that folded and faulted the early Tertiary rocks. Time-equivalent plateau lavas covering a similar area to that of the Chilcotin Group occur over the Columbia Plateau in neighbouring Washington State. Curiously, the American basalts are world famous but few

geologists outside of British Columbia have heard of the Chilcotin Basalts!

In the Kelowna area, where streams and glaciers have eroded through the plateau basalt sequence, a conspicuous, steep scarp has developed (Figure 2-3). Up to thirty individual basaltic flows occur near Kelowna in the area of Daves Creek. Individual flows are thin, and the total stratigraphic thickness of the basaltic flows rarely exceeds 100 metres. One of the most striking sections through the basalts occurs along the West Kettle River valley just south of the Big White access road (photo on back cover). At least three basalt lava flows are exposed for over one kilometre along a forestry access road (West Kettle Columns in Figure 5-12). These basalts display spectacular pillar-like structures called columnar joints which develop when lava shrinks and fractures into vertical columns during cooling. They overlie a sand and gravel deposit, an example of the stream activity just prior to the plateau basalt event.

Figure 8-4: A southerly view from the Kelowna Floating Bridge shows an escarpment along Lakeview Heights which is formed from the Lambly Creek valley basalt.

A VALLEY BASALT EVENT

The last volcanic event to affect the Kelowna area is only 762,000 years old (Middle Pleistocene age). It is a valley basalt called the Lambly Creek basalt. It flowed twenty-five kilometres from a source north of Lambly Lake, down Lambly (Bear) Creek, and into the Okanagan Valley (Figure 1-4). It is preserved as a lava bench on the west side of the lake near the Okanagan Bridge and below the Lakeview Heights residential area, here referred to as the Lakeview Escarpment (Figures 8-4 and 9-4).

This valley basalt gives some idea of the topography in the area three quarters of a million years ago, and supports the idea that at least some of the present tributary valleys were partly developed by that time.

The Lambly Creek basalt shows chemical similarities to the Chilcotin Group basalts. This reminds us that Chilcotin volcanic activity has probably not ended in the interior of British Columbia. Fortunately, the probability of an eruption during our life time appears extremely small.

Figure 9-4: This is a view of vesicular basalt (762,000 years old) which makes up the Lambly Creek basalt forming Lakeview Heights Escarpment.

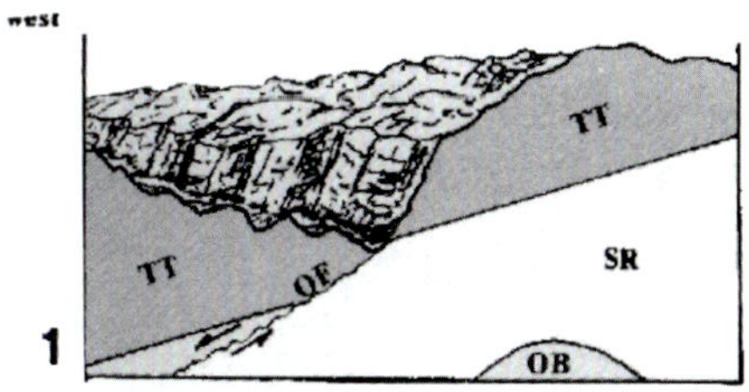

1

Okanagan Fault ("OF") is activated and starts to move the Shuswap rocks ("SR") toward the surface creating a rugged scarp to the east over 60 million years ago. The top layer of rocks represents transported terrane that may have initially formed on the side of the continent ("TT"). Granite of the Okanagan Batholith ("OB") formed 170 million years ago.

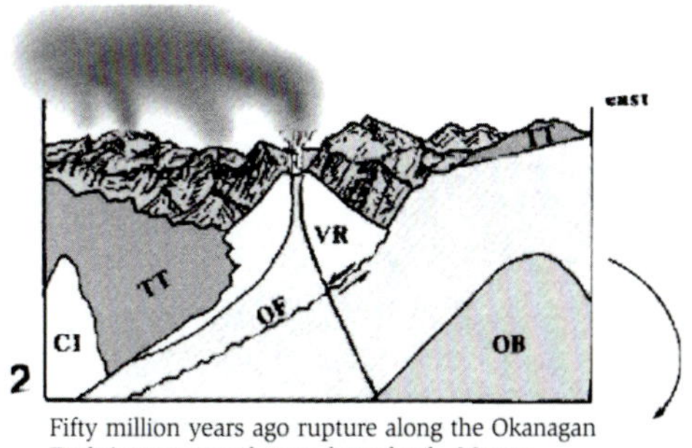

2

Fifty million years ago rupture along the Okanagan Fault increases and extends to depth. Magma chambers form and feed extensive volcanism on the surface ("VR") near the present site of Kelowna (Kelowna Caldera) and also create Coryell Intrusions ("CI") in the subsurface about 50 million years ago. The Monashee Gneiss rocks continue to rise as the terrane to the west slides off.

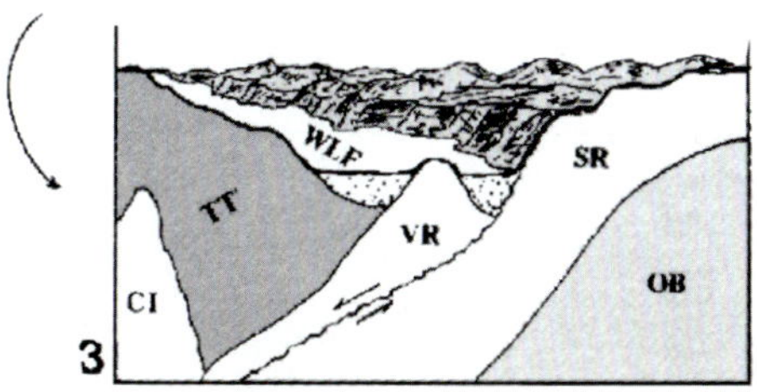

3

Volcanism ceases and a large river develops burying the rough landscape of the volcanic rift valley. These rocks are represented by rocks of the White Lake Formation ("WLF"), composed of sandstone, conglomerate and thin coal beds that outcrop in the Kelowna area.

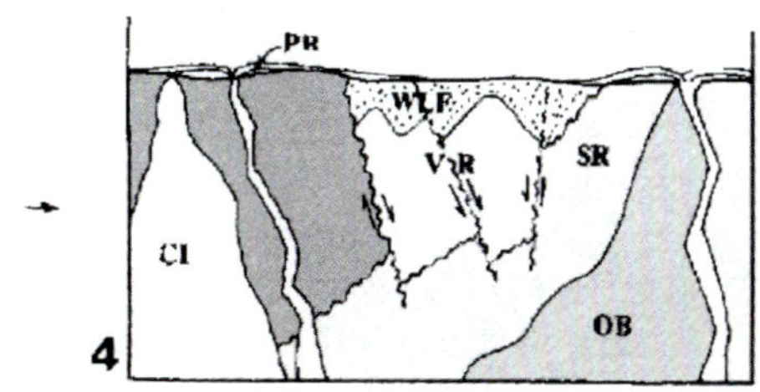

4

The region is uplifted, faulted, folded and subjected to extensive erosion creating a vast rolling upland known as the Thompson Plateau. Between ten and twenty million years ago deep cracks develop and allow the periodic escape of highly fluid lava that forms plateau basalt ("PB") belonging to the Chilcotin Group.

Figure 10-4: Summary of Tertiary Geologic History at Kelowna.

SUMMARY

It is clear from this brief description of the tectonic and volcanic history of the Kelowna area that the interior part of British Columbia has been geologically very active over the past fifty million years. Total uplift in the Coast Range Mountains to the west is on the order of four kilometres in the last ten million years. It follows that there was extensive erosion in the region during this prolonged period of uplift. Erosion is one reason why there are many gaps in the geologic record in the Okanagan area, compared to other areas of the world. A sequence of diagrams in Figure 10-4 summarizes the main events.

Figure 11-4: A portion of a map drawn by J.B. Landers and J.W. Trutch in 1871 showing the Roman Catholic Mission at present day Kelowna and various trails including the Brigade Trail.

CHAPTER 5

THE ICE AGE IN THE OKANAGAN

by Murray A. Roed

The continued uplift of the region and the initial erosion of the present valleys down through the adjacent uplands to the end of Tertiary time combined with a gradual cooling of the climate set the stage for another dramatic geologic event, the Ice Age. This is the Pleistocene Epoch and it began approximately 1.6 million years ago. At first ice began accumulating at topographic high points and began to flow under the influence of gravity. The tongues of ice filled the valleys and coalesced in the larger valleys. Finally ice overtopped the valleys and spread out onto highlands. Glacial build-up reached its maximum when all of the ice accumulation areas merged. At this time, the Cordilleran Ice Sheet was formed and covered the entire province. From space, it probably looked like a vast brilliant blueish and white lobate collage marked by widely spaced, gently sloping dome-like mounds of regional ice accumulation centres (Figure 1-5). Only the very highest peaks projected above the ice as nunataks. To the east, along the Rocky Mountain Foothills in Alberta, the Athabasca lobe of the Cordilleran Ice Sheet merged with the Continental Ice Sheet and they flowed south together for several hundred kilometres into the United States (Roed, 1967). In the west, ice flowed out of the Cordillera, traversed the Continental Shelf and calved into deep water of the Pacific Ocean. Most of the ice moving through the Okanagan originated in a major accumulation zone about 800 kilometres (500 miles) to the north in the northern Selkirk Mountains. Glacial ice flowing over the southern interior terminated in broad lobes that covered large parts of northern Idaho, Montana and Washington. These lobes are here referred to as the Columbia, Kootenay, Okanagan and Fraser lobes. The ice sheet was 1000 to 3000 metres thick over much of the Interior of British Columbia. The most recent glacier at Kelowna may have been 3000 metres thick along the centre of the valley of Okanagan Lake. It flowed to the south and southeast as indicated by striae on bedrock in the area. Large boulders here and there known as erratics (Figure 2-5) attest to erosional power of the ice.

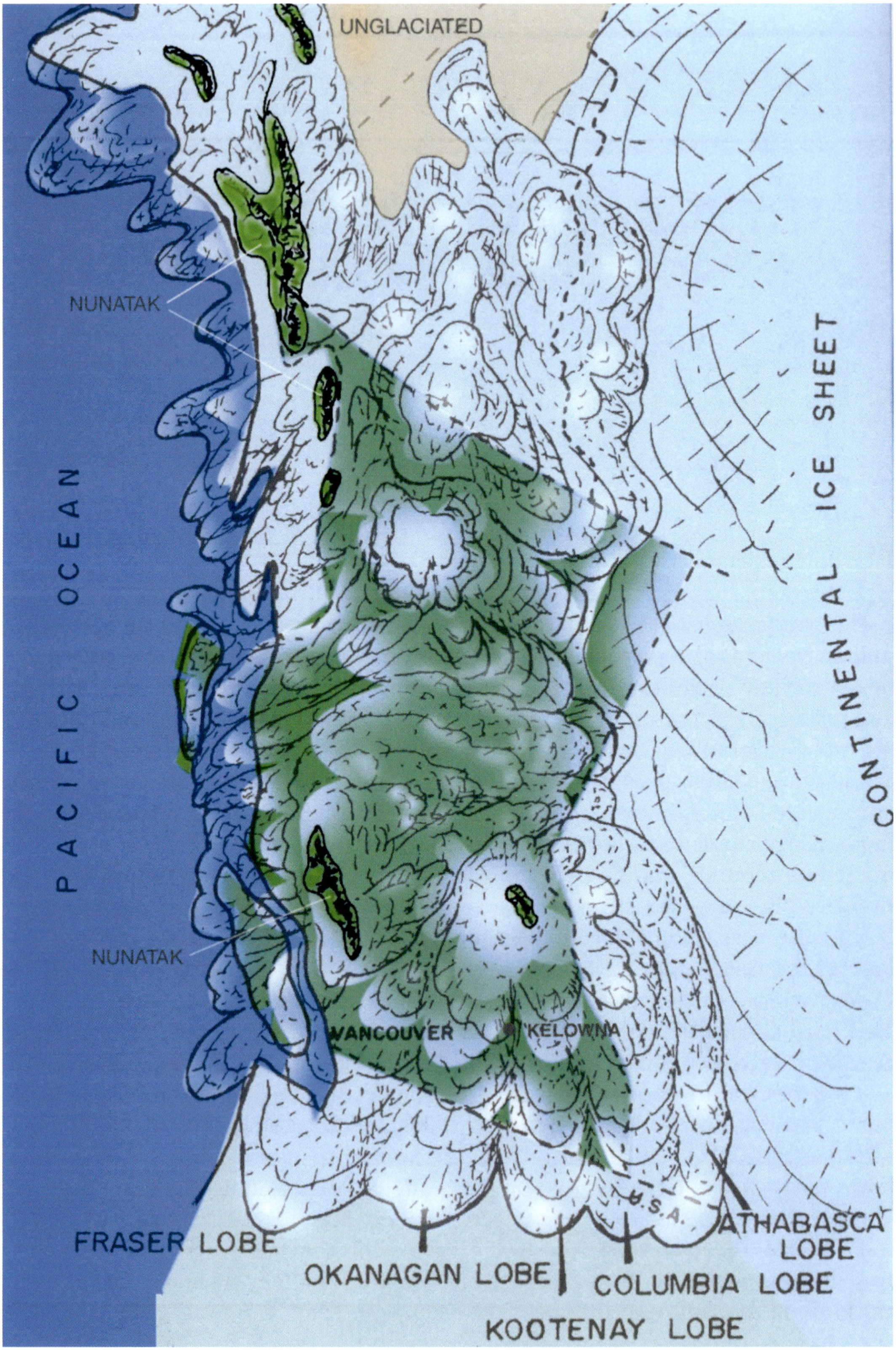

Figure 1-5. An "artistic" version of the Cordilleran Ice sheet from space.

MULTIPLE GLACIATIONS

There have been at least four intervals of Cordilleran ice build-up each separated by an interglacial period when the climate was warm and ice melted. However, according to Don Easterbrook, working in the State of Washington, at least six glaciations of the Cordilleran Ice Sheet entered the state from British Columbia during the Pleistocene Epoch. Canadian geoscientists (Fulton et al, 1992) have uncovered evidence of similar multiple glaciations at a locality near Merritt. Generally, in British Columbia, only the last two glaciations are known in any detail. The older is referred to as the Okanagan Centre Glaciation, and the younger as the Fraser Glaciation. See the Addendum that documents a much older glaciation.

Figure 2-5: This large boulder of gneiss in Mission Creek valley is an example of an erratic. The rock was plucked from bedrock by a glacier and carried to its present spot where it was deposited when the ice melted.

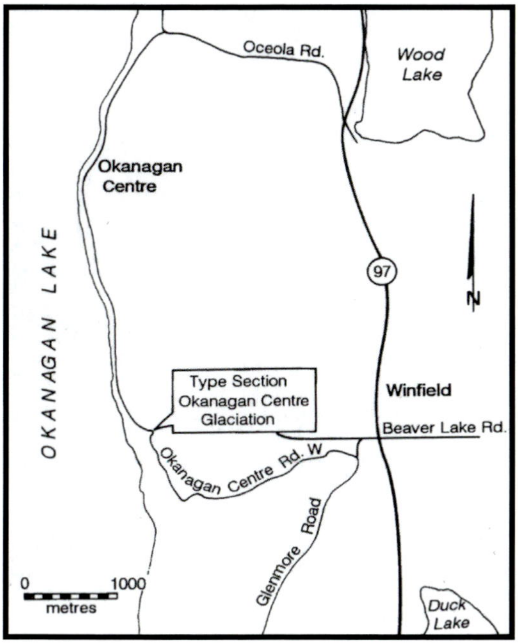

Figure 3-5. Location of the Type Section of Okanagan Centre Glaciation.

OKANAGAN CENTRE GLACIATION

Two and one half kilometres south of Okanagan Centre (Figure 3-5) a sequence of glacial and nonglacial sediments ninety-eight metres thick is exposed in a gully (Figure 4-5) on the east side of the road. Fulton and Smith (1978) of the Geological Survey of Canada determined that they represent the last two major glacial advances (25,000 and 50,000 years ago) separated by a non-glacial interval, and an older Interglacial interval (100,000 years old).

The deposit that is laid down directly when glacier ice melts is called till (Figure 6-9). The till of the Okanagan Centre Glaciation is underlain by a series of glacial lake deposits, sand, gravel of glacial and fluvial origin and

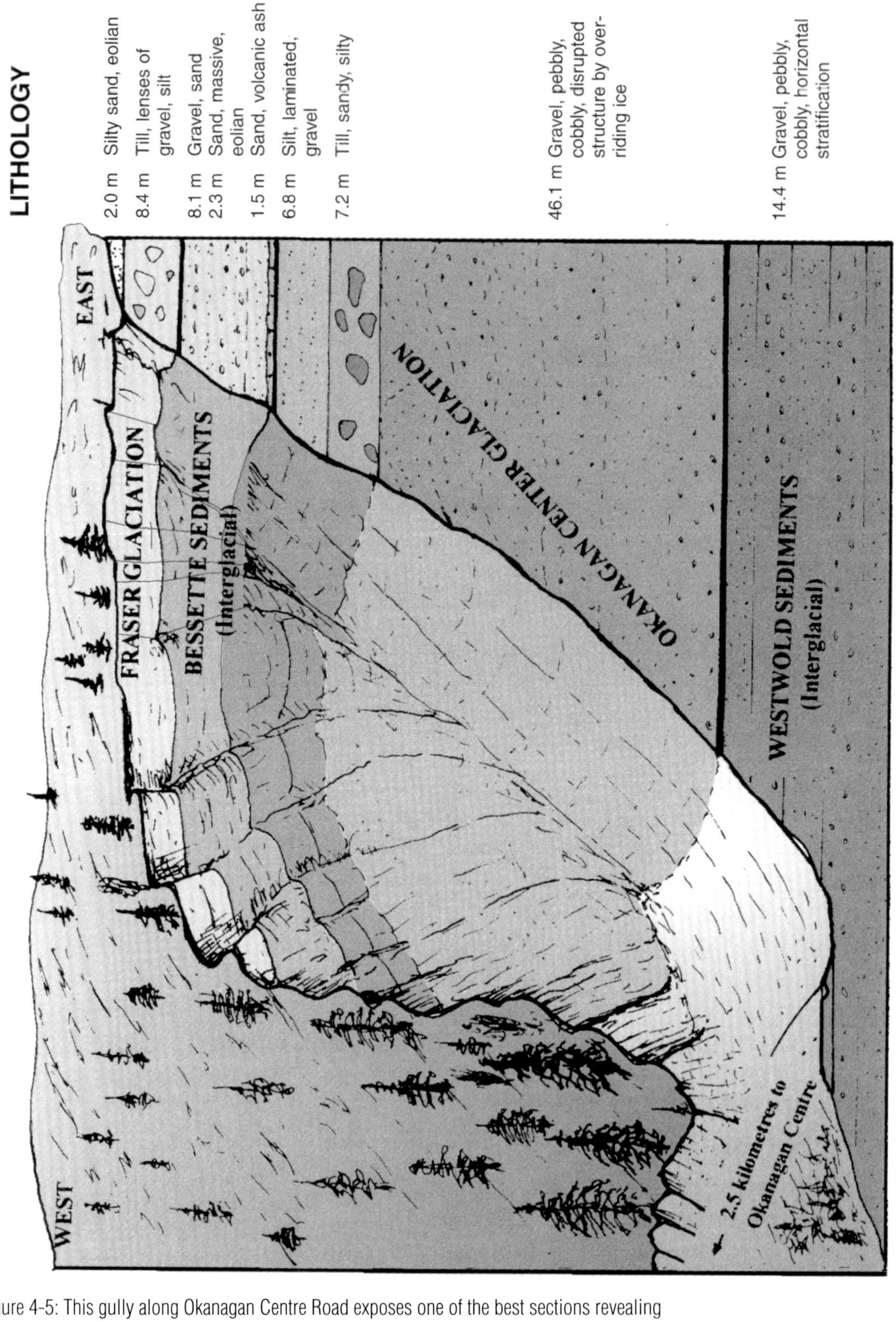

Figure 4-5: This gully along Okanagan Centre Road exposes one of the best sections revealing the variety of glacial deposits formed in the last 100,000 years in British Columbia.

beach deposits referred to as the Westwold Sediments. Fossils recovered from these deposits at Westwold, B.C., include plant fragments, mollusk shells, bison bones, fish, beetles, and rodents suggesting a warm climate (Fulton and Smith, 1978) during a 100,000 year old Interglacial period in the Okanagan.

The till of the Okanagan Centre Glaciation is overlain by another series of sediments representing a regional period of nonglacial activity referred to as the Olympia Nonglacial Interval. These sediments were formed in many valleys after the ice of the Okanagan Centre Glaciation had melted. They are known as the Bessette Sediments and are exposed in Bessette Creek near Lumby where they are 22 metres thick and consist of interbedded fluvial gravel, sand and silt containing plant remains and at least two layers of volcanic ash. These deposits are overlain by till and sediment related to the most recent Cordilleran glaciation called the Fraser Glaciation.

FRASER GLACIATION

The best known glaciation in British Columbia is the Fraser Glaciation because its deposits form much of the present surface terrain of the Province. In many localities only one till, representing one ice advance, occurs but in some places at least two tills occur which means that there were at least two advances of the Fraser Glacier. The Fraser Glacier completely engulfed British Columbia as recently as 17,500 years ago following an ice build-up that began about 25,000 years ago due to climatic change.

Fraser Glaciation deposits consist of till that mantles most of the terrain today, outwash deposits of various sorts, late Pleistocene-earliest Holocene glacial lake deposits, and areas of bare bedrock that have been molded or scraped by ice, or eroded by meltwater. A variety of these deposits and the landforms produced occur near the major centres of Vernon, Kelowna and Penticton as summarized in the surficial geology map of Figure 5-5.

GLACIAL LAKE PENTICTON

Approximately 13,000 years ago the retreating lobe of the Fraser Glacier that occupied the Okanagan Valley formed an ice dam near Okanagan Falls south of Penticton. Meltwater flowing into the valley ponded behind this dam. This formed a lake much larger and deeper than Okanagan Lake known as Glacial Lake Penticton. Sediments deposited in this lake basin form the high, silt bluffs up and down the valley. These silt bluffs and associated terraces impart much of the beauty to the landscape for which the Okanagan is famous. Glacial Lake Penticton reached a maximum elevation of 457 metres, over 100 metres above the present lake level. The lake extended as far north as Enderby, probably joining a similar lake in the Thompson River valley, and completely submerging the present sites of Penticton, Kelowna, Vernon and Armstrong.

The Okanagan glacier lobe persisted for some time, melting very slowly as the water in Glacial Lake Penticton built up. Finally, as the core of stagnant ice melted, and as the ice dam periodically was breached, the level of the glacial lake began to drop. A series of terraces (silt bluffs) and erosional scarps in the Kelowna area were formed as the lake drained in distinct intervals (Figure 1-14). The last bit of glacial ice persisted so long in the valley that when it did finally melt, a depression was left that filled with water and formed the present Okanagan Lake. Thus, Okanagan Lake

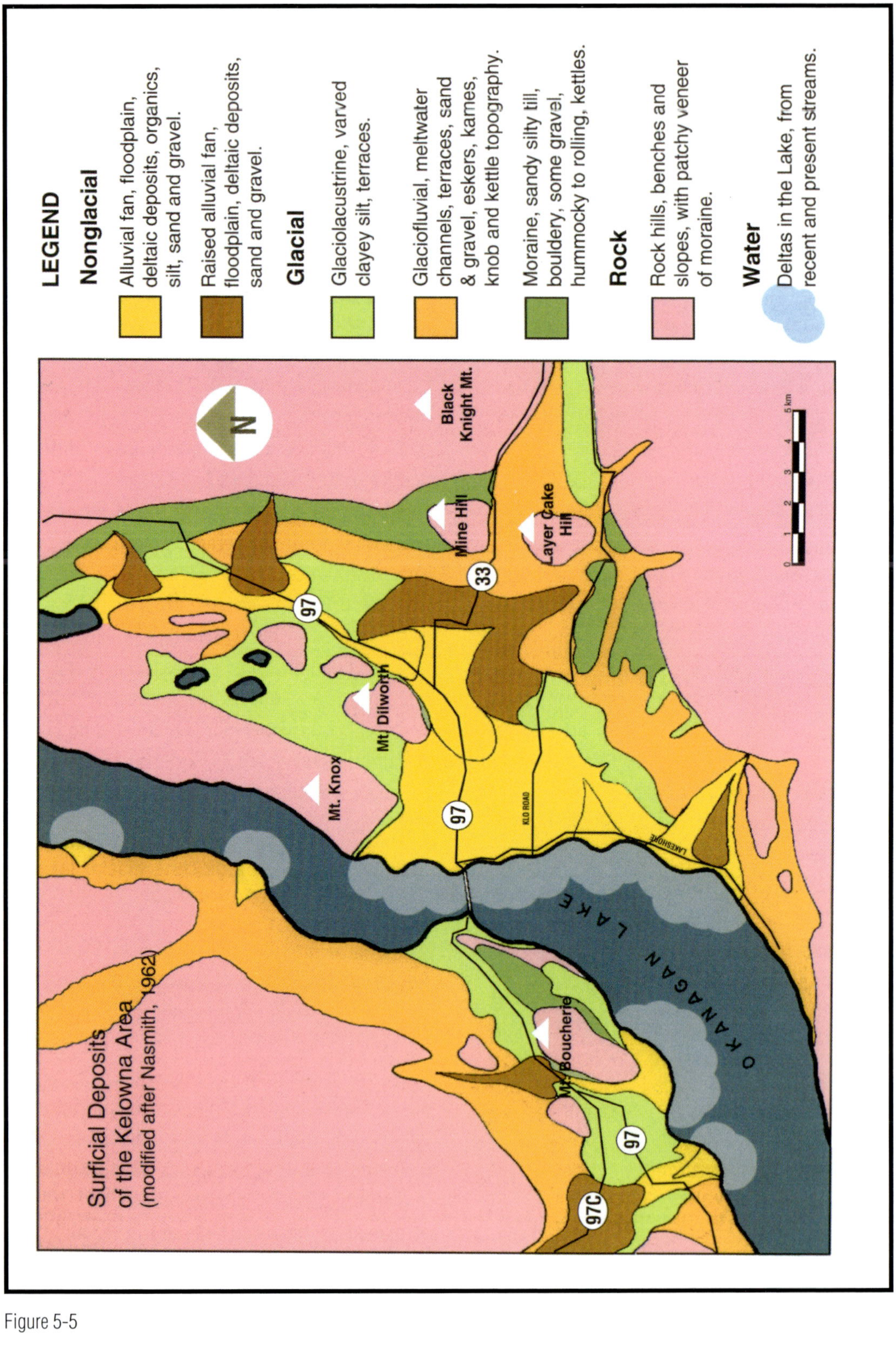

Surficial Deposits
of the Kelowna Area
(modified after Nasmith, 1962)
LEGEND
Nonglacial
Alluvial fan, floodplain, deltaic deposits, organics, silt, sand and gravel.
Raised alluvial fan, floodplain, deltaic deposits, sand and gravel.
Glacial
Glaciolacustrine, varved clayey silt, terraces.
Glaciofluvial, meltwater channels, terraces, sand & gravel, eskers, kames, knob and kettle topography.
Moraine, sandy silty till, bouldery, some gravel, hummocky to rolling, kettles.
Rock
Rock hills, benches and slopes, with patchy veneer of moraine.
Water
Deltas in the Lake, from recent and present streams.
N
Black Knight Mt.
Mine Hill
Layer Cake Hill
Mt. Dilworth
Mt. Knox
Mt. Boucherie
KLO ROAD
LAKESHORE
OKANAGAN LAKE
97
33
97C
0 1 2 3 4 5 km

Figure 5-5

originated partly as a "kettle lake". A kettle is a depression in the ground where part of a glacier became buried. When the ice melts, the ground collapses and forms the depression called a kettle, or pit. If the kettle contains water, it is referred to as a kettle lake.

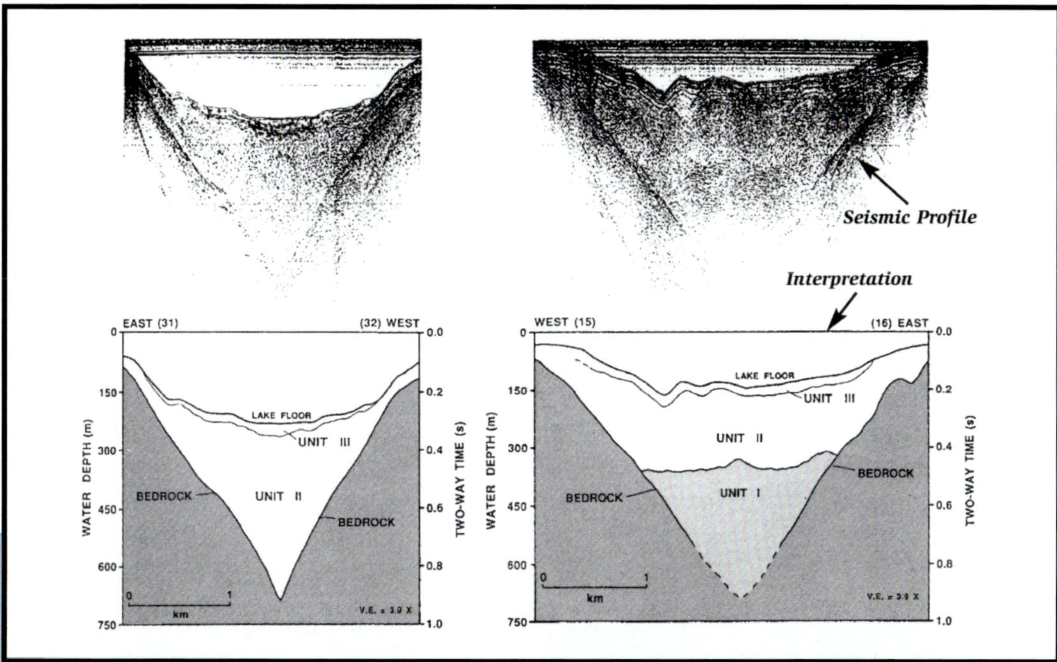

Figure 6-5: Interpreted profiles (5X vertical exageration) below Okanagan Lake from seismic studies (from Eyles et al, 1990). The left profile is just north of Bear Creek; the profile on the right is just south of Squally Point. See Figure 7-AD for a new interpretation of these cross-sections.

THE BURIED OKANAGAN VALLEY

It was not until long after the Plateau Volcanic event during Miocene-Pliocene time that the ancestral drainage of the present Okanagan landscape began to form. Due to structural weakness along parts of the Okanagan Valley Fault zone, the Okanagan Valley became a major stream valley, tributary to a large river system that drained south into the United States. By the time the climate had cooled near the end of the Tertiary Period, the region was still an upland which was slowly becoming dissected by the new "preglacial" river system. Most of the major valleys in the interior likely originate from this time and may have been interconnected.

During some of the glacial build-ups, ice over three kilometres thick accumu lated above the deepest valleys. Ice this thick has tremendous scouring power near its base and sides. This explains the astounding overdeepening that affected the Okanagan Valley. The bedrock below Okanagan Lake is as much as 640 metres below sea level! The surface of the lake is 342 metres above sea level.

Our knowledge of features below the lake was enhanced by a seismic reflection survey conducted in October of 1989 (Eyles et al,

1990). Profiles obtained in this work are shown in Figure 6-5. The 5X exageration depicts a false V-shape of the bedrock valley (see Addendum). It is apparent also that up to 750 metres of material has accumulated in the bedrock valley of the lake. Some of this is glacial, glaciolacustrine and interglacial in origin, and some is related to sedimentation by streams that now empty into the lake.

The overall relief on the bedrock surface in today's landscape at Kelowna is 2810 metres (the difference in elevation between the deepest bedrock at the bottom of the lake, -640 metres, and that of neighbouring Little White Mountain, 2170 metres). This exceeds the relief of the Grand Canyon in Arizona at 1600 metres.

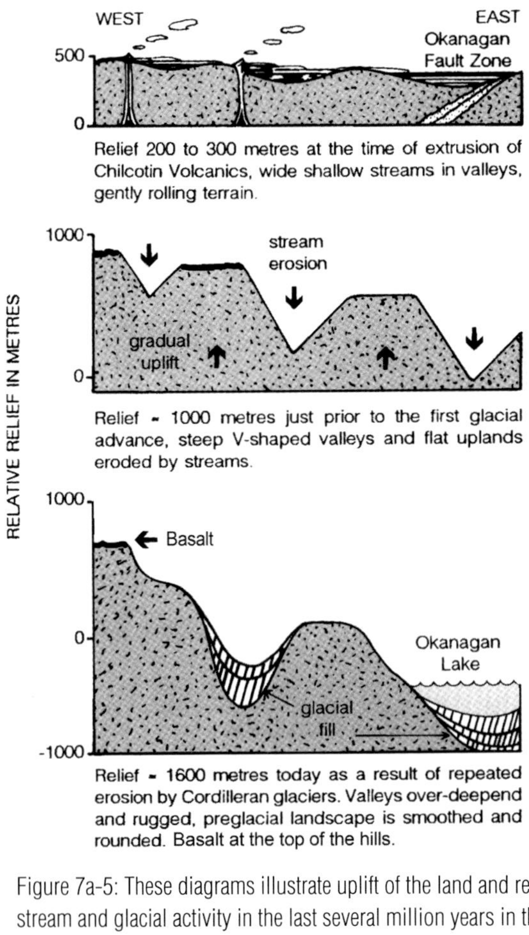

Figure 7a-5: These diagrams illustrate uplift of the land and resulting erosion from a combination of stream and glacial activity in the last several million years in the Okanagan Valley.

OKANAGAN LAKE TODAY

Okanagan Lake (Figure 7b-5) is approximately 120 kilometres long and averages 3.5 kilometres wide. The watershed of the lake is 6178 square kilometres, about one third the size of Switzerland. The lake exists today because it is a depression carved out by repeated glaciations. For this reason Nasmith (1962) called the lake a "fjiord lake" resembling the spectacular fjiords of Norway. Its deepest point is 232 metres off of Whisky Island, north of Okanagan Centre.

Although deep throughout most of its length, there are numerous terrace-like submarine platforms that are less than 10 metres deep, shallower parts of which are definite marine hazards. The most extensive "terrace" stretches from Lakeview to Westbank on the western side of the lake. During low water it is possible to walk into the lake over 100 metres along Manhattan Beach, at Poplar Point (both in the north part of Kelowna) and at the sites of some deltas, for example, Mill Creek, Mission Creek and Bellevue Creek. Figure 5-5 shows the locations of numerous deltas that have built out into the lake. On the other hand it is not unusual for the lake to be over 100 metres deep a mere 10 to 20 metres from shore along steep bedrock outcrops.

Detailed examination of original bathymetric data (Canadian Hydrographic Service) reveals a substantial complex of ridges and depressions in the deep parts of Okanagan Lake. The features are curvalinear and trend north-south. Some depressions are closed. The closed depressions are likely pits and may represent former positions of glacier ice. The submarine topography of the lake requires further study, and the same for the sedimentary fill in the lake basin. Over 90 cubic kilometres of sediment alone has been deposited in the lake.

Elements of the hydrologic cycle of the lake are given in Chapter 13. The level of the lake is now controlled by dams at Penticton, Okanagan Falls and Vaseux Lake and on a yearly basis fluctuates about 1.5 metres. 2004 has been cited as the lowest lake level in 100 years! It has been estimated that water in the lake has a "residence" time of about 60 years.

Figure 7b-5: Okanagan Lake valley looking north from the Knox Mountain viewpoint.

PRESENT DAY KELOWNA

The landscape and main geologic materials as they exist today are shown in a cross-section of the geology of the valley in Figure 4-4, page 32. It represents what you might find if the valley was cut open from near Mount Boucherie on the west, through the lake and subsurface, to below Little White Mountain in the east.

On the west side of the lake is a fine example of a remnant upland (Hayman/MacDougall Upland) of the Thompson Plateau, capped with Plateau basalt at Carrot Mountain in the west. Proceeding east towards the lake, the cross-section shows the subsurface extent and volcanic fill of the Kelowna Caldera Basin. Note the faults affecting the subsurface. The Coryell intrusion which outcrops in Trepanier Creek and along the Highway 97C Connector represents a magma chamber that fed Tertiary volcanoes (e.g. Figure 8-5). Sandstones and conglomerates of the White Lake Formation partly filled the pre-existing rugged topography of the rift-like valley. Lake Okanagan shows a thick infilling of glacial and post-glacial sediments, and deep fiord-like erosion of its bedrock base. East of the lake is the metamorphic Monashee Gneiss complex and the bounding Okanagan Valley Fault zone. Arrows indicate the direction of former movement along the fault. Further to the east Little White Mountain is underlain by granitic intrusive rocks of the Jurassic-Cretaceous age Okanagan Batholith, with remnants of the intruded Monashee Gneiss.

Figure 8-5: Feeder Dikes, Westbank. This photo shows two feeder dikes to volcanic rocks in the vicinity of Shannon Lake Golf Course near Westbank. The dikes have cut through older rocks including a black shale that can be seen in the photo. The dikes are varieties of a porphyritic andesite with orthoclase crystals easily visible.
John D. Greenough, photo

Figure 9-5: Acrylic painting of Fintry Falls in Fintry Provincial Park, by M. Roed. The falls have eroded through hard granite of the Nelson Plutonic Complex. Numerous dikes occur at the site and the rocks are heavily fractured.

CHAPTER 6

CLIMATIC CHANGE, THE LAST FIFTEEN THOUSAND YEARS IN THE OKANAGAN

by Ian R. Walker

The glacial maximum of the last ice age occurred approximately 16,000 BC, after which warming began. It took about six thousand years to finally melt the Fraser Glacier. The reflective surface of the ice had helped maintain a cold glacial climate; however, as the ice melted, the dark freshly exposed soils and bedrock absorbed heat and accelerated the warming trend. Cool air overlying the stagnant ice rushed down the glacier, replacing the air warmed by exposed sediment and bedrock and generating strong down-valley winds.

Ice melted quickly off the adjacent Okanagan highlands as indicated by radiocarbon dates on the earliest sediments deposited in high-elevation lakes. The deeply eroded canyons (e.g. Myra Canyon) of creeks at higher elevations may have formed as a result of powerful rushes of meltwater coming off the highlands during initial melting of Fraser ice, probably between 12,000 to 10,000 BC.

Much of the eroded material was likely deposited on the surface of the stagnant glacier occupying the lower portion of the valley. The highlands were probably largely ice-free by 11,000 to 10,000 BC.

The ancestral Okanagan River was dammed by stagnant ice near Vaseux Lake, creating a large glacial lake known as Glacial Lake Penticton. At its maximum extent Glacial Lake Penticton extended from Okanagan Falls north beyond Vernon, and encompassed Skaha, Wood, Kalamalka and Swan lakes, in addition to Okanagan Lake. By 10,000 BC glacier ice had largely disappeared, but the level of Glacial Lake Penticton was still well above the current level of Okanagan Lake. Land north of Okanagan Falls had been depressed by the great weight of ice. With the weight removed, the land began to rebound promoting drainage toward the south, and a lowering of lake level.

Freshwater fish species (e.g., large scale sucker, chiselmouth, redside shiners, squawfish and leopard dace) which survived glaciation south of British Columbia in the Columbia River and its tributaries were likely able to disperse northward via meltwater connections from Glacial Lake Penticton into the Thompson River, and hence, throughout the Fraser River system, and from there into adjacent watersheds. Minute larvae (glochidia) of freshwater mussels clung to these fish, and spread their own species northward.

A radiocarbon date from the north Okanagan suggests that this meltwater connection had disappeared by about 9000 BC. A date from the bottom of the former Kelowna bog (just south of the Kelowna airport) also indicates that Glacial Lake Penticton had largely drained by 8,000 BC. Much of Kelowna was probably an extensive marshy delta. The rest of the world was just emerging from the Stone Age (Figure 1-6).

2000 AD

PRESENT DAY - European settlement, warming trend, retreat and dissappearance of alpine glaciers.

LATE HOLOCENE (NEOGLACIAL)

LITTLE ICE AGE - cool climate, glacial advances in mountains of southern BC

MEDIEVAL WARM PERIOD - Vikings establish settlement at L'Anse aux Meadows, Newfoundland

BRIDGE RIVER ASH - Mount Meagher, near Pemberton, explodes ca. 500 BC

TIEDEMANN ADVANCE - cool moist climate, glaciers advance in Coast Mountains and elsewhere; hemlock spreads rapidly through Kootenays

MOUNT ST. HELENS Y ASH - ca. 1600 BC

2000 BC

MIDDLE HOLOCENE (MESOPTHERMIC)

MID-HOLOCENE COOLING - forests encroach on Okanagan Valley grasslands; cedar spreads rapidly northward in coastal BC; coastal First Nations wood-working culture appears

MOUNT MAZAMA ASH - colossal eruption destroys Mount Mazama in Oregon (Crater Lake), raining ash across western North America

6000 BC

EARLY HOLOCENE (XEROTHERMIC)

XEROTHERMIC - warm, dry summers, and frequent fires promote maximum expansion of valley grasslands; forests restricted to moist riprian areas or higher elevations; saline lakes develop in former freshwater basins; winters may have been colder than today

PLEISTOCENE/HOLOCENE TRANSITION - abrupt shift from glacial to interglacial climate; ice rapidly disappears; forests replace tundra & cold-steppe; meltwater from Thompson River is diverted by ice dams into the Okanagan; Glacial Lake Penticton gradually drains to south as land rebounds

10000 BC

PLEISTOCENE (LATE-GLACIAL)

LATE-GLACIAL - ice sheet thins, exposing Okanagan highlands; tundra or cold-steppe vegetation colonises freshly-exposed surfaces

Figure 1-6. Geologic column illustrating major events since the last glaciation. Note: All dates are provided in calendar years. Dates in older publications were commonly provided in "radiocarbon years". Without correction, the "radiocarbon" and "calendar" dates may differ by up to 2000 or more years.

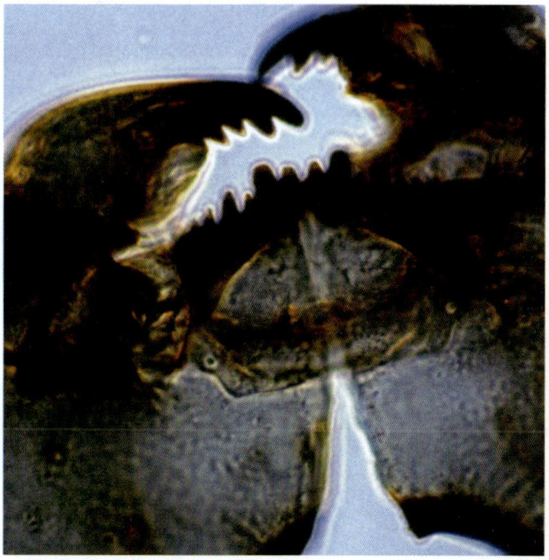

Figure 2-6: Fossil head capsule of Tribelos, a chironomid or non-biting midge.
(I.R. Walker photo)

AFTER THE LCE MELTED

Scientists have reconstructed the vegetation and associated climatic changes in the Okanagan region from fossil pollen (Alley, 1976), and plant and insect fragments. The head capsules of minute larval flies (chironomids, or non-biting midges) have proven especially useful (Figure 2-6). The derived summer temperature record (Figure 3-6) reveals cold conditions before 10,000 BC, rapid postglacial warming, maximum postglacial summer temperatures ca. 9600 to 6000 BC, a gradual decrease in summer temperatures from ca. 6000 to 2000 BC, and a relatively stable climate over the past 4 millennia.

LATE-GLACIAL (11,000 TO 10,000 BC)

Lake sediment cores from the south Okanagan highlands and vicinity (Mount Kobau, Crater Mountain, Cathedral Provincial Park) studied by M. Heinrichs preserve pollen of tundra or cold-steppe species (e.g., sage, grasses, juniper) that initially colonized the freshly deglaciated highlands. It was probably at this time that many now-abandoned gullies formed high on Okanagan Valley slopes. The cold and dry conditions delayed the establishment of forests, but summer temperatures increased rapidly ca. 10,000 BC, at the beginning of our current interglacial warm period, the Holocene.

EARLY HOLOCENE (XEROTHERMIC)

As Glacial Lake Penticton drained, streams eroded deeply through the benches surrounding the valley, and in depositing their sediments built broad alluvial fans into the lakes. Much of Kelowna is built on the alluvial fan derived from Mission and Mill creeks. In the south Okanagan, alluvial fans from Penticton, Ellis and Shingle creeks merged to create the broad flat that now separates Okanagan and Skaha lakes. In the north, the same process built Fintry delta, and at the mouth of Upper Vernon Creek, the flat in Lake Country separating Duck (Ellison) and Wood lakes.

The arrival of the first trees in the region ca. 9000 to 10,000 BC is recorded not only by their pollen (Figure 4-6), but by cone scales and needles of lodgepole pine, Engelmann spruce, and subalpine fir which are preserved in sediments of Okanagan highland lakes. As the land surface was stabilized by forests, glacial lake deposits of inorganic silts and clays were buried by organic-rich sediments, derived from the decay of plants and animals of aquatic and terrestrial origin. M. Heinrichs showed that the warm temperatures brought drought. Many small low-elevation lakes initially overflowed with fresh water but with increased evaporation, they became increasingly shallow. As the remains of minute algae (diatoms) and midges indicate, many small Okanagan Valley lakes became highly saline – like Robert Lake, the shallow playa west of the North Kelowna, university campus. Salts precipitated from the water, forming white encrustations of evaporite minerals (salts) along the lake shores.

In recognition of the drought, scientists commonly refer to this period as the Xerothermic interval. Although warm summer temperatures allowed forests to quickly establish, the hot dry summers restricted tree growth. Grasslands were more extensive, and at higher elevations the forests were more open than today. Frequent fires helped to maintain this open landscape.

There is abundant evidence that summer temperatures were warmer, but winter temperatures may actually have been colder. The impact of this greater seasonality is difficult to discern, except by comparison to modern conditions. For example, we can imagine that if such cold winters prevailed today, the Okanagan climate would be unsuitable for vineyards and many tree-fruits, whereas pine beetle outbreaks would likely be much less severe.

MIDDLE HOLOCENE (MESOTHERMIC)

About 5700 BC, a colossal volcanic eruption shook Oregon, destroying Mount Mazama

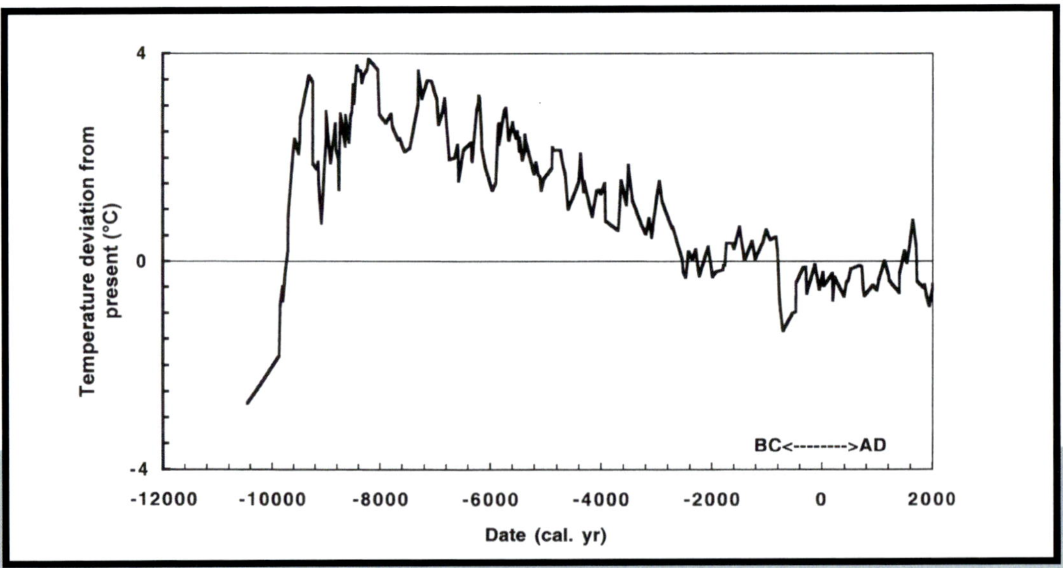

Figure 3-6. Midge derived palaeotemperature reconstruction for southern interior British Columbia (Adapted from Rosenberg et al., 2004 and Walker & Pellatt 2003).

Figure 4-6. Fossil pine pollen grain. (S. Rosenberg photo)

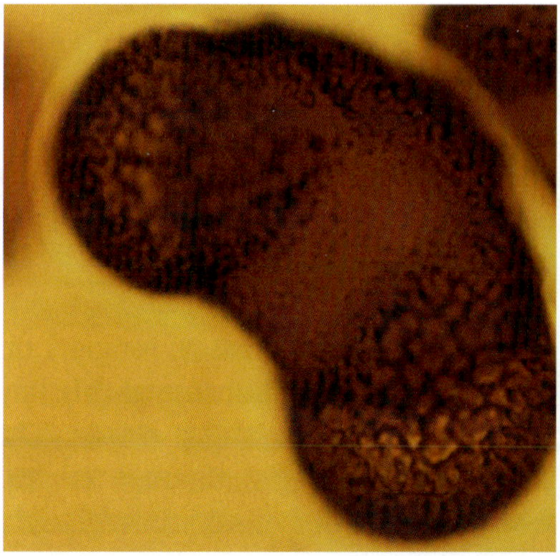

and creating Crater Lake. The tephra (volcanic ash) from this eruption blanketed much of western North America, including the Okanagan Valley. In Okanagan Valley lakes, layers of this ash are up to 60 cm thick, and thin traces occur as far away as the Greenland ice sheet.

This spectacular eruption marks the onset of the middle Holocene in southern British Columbia, and the transition from the hot dry summer environment to our modern climate. This period of transition and intermediate temperatures, from about 6000 BC to 2000 BC, is referred to in British Columbia as the "Mesothermic interval".

At the beginning of the Mesothermic, extensive grasslands blanketed the valley slopes, and forests may have been largely confined to highlands, except for the alluvial cottonwood fringe tracing stream courses to valley lakes. Grassland species such as antelope brush that are represented only by small isolated relict occurrences in the Kelowna area today, were likely more abundant early in the Mesothermic interval. However, as summer temperatures decreased (Figure 3-6), the soils became more moist, and forests advanced downslope into the grasslands. In coastal areas of British Columbia a characteristic tree of the temperate rain forest, western red cedar, expanded its range northward, closely followed by the wood-working tradition of coastal first nations. Evidence of the Mid-Holocene cooling trend is also evident in the Garibaldi phase glacial advances of the Coast Mountains, dating between ca. 4000 to 5000 BC.

LATE HOLOCENE (NEOGLACIAL)

Since 2000 BC the climate and vegetation of the Okanagan Valley has changed little. The moister conditions and more exuberant plant growth have stabilized most sand dunes. Elsewhere, in the alpine areas there are signs of renewed glacial activity. Several glacial advances in the Coast range have been dated between 1000 BC and 100 AD. Advances of the Berendon (ca. 1000 BC to 300 BC), Frank Mackie (ca. 800 BC), Tiedemann (ca. 400 BC), and Gilbert (ca. 100 AD) glaciers are collectively referred to as the Tiedemann Advance. The Battle Mountain advance (ca. 1700 to 400 BC) in the Shuswap Highlands, and the Neoglacial maximum of the Bugaboo glacier (ca. 600 BC) also date approximately to this time. To the east the cooler and moister conditions allowed hemlock to spread rapidly throughout the Columbia Mountains after 2500 BC.

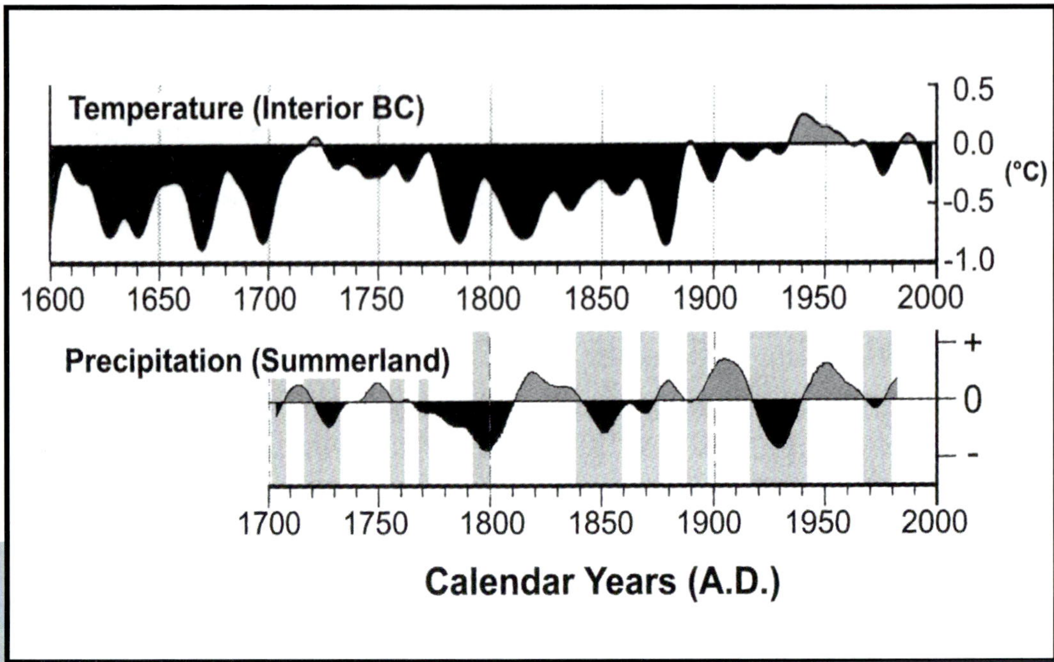

Figure 5-6. Tree-ring based reconstructions of recent Okanagan region temperature (1600-1997 AD) and precipitation (1704 - 1980 AD) changes. Adapted from Wilson & Luckman (2003) and Watson & Luckman (in press).

Between 1100 and 1900 AD (referred to as the Little Ice Age), many glaciers advanced in British Columbia and throughout the northern hemisphere. The climate variation through much of this interval is recorded in the ring-widths of Okanagan region trees (Figure 5-6). Narrow rings in Engelmann spruce near treeline on Big White record cold years, whereas broad rings indicate warmer times. Similarly, wet vs. dry conditions are read from the ring widths of Douglas fir and ponderosa pine growing near the valley bottom. Cold temperatures are evident through much of the 17th and 19th centuries Little Ice Age, as compared to the 18th and 20th centuries. A prolonged drought occurred from about 1770 to 1810 AD, as well as the historically recorded drought of the 1930s.

THE PAST CENTURY

The environmental changes of the past 100 years reflect both natural events and human activities. For example, metals contained in the sediments of Wood Lake reveal the impact of European settlement and subsequent agricultural and industrial development on the lake (Figure 6-6). Peaks in lead and arsenic record the use of lead arsenate as a pesticide for codling moth control prior to 1945. Recent increases in molybdenum reflect the deteriorating oxygen conditions at Wood Lake, culminating with the lake's prolific algal blooms of the late 1960s. The increased calcium indicates that calcium carbonate crystallized around these algae and settled to the lake floor. Gradual increases in lead and mercury respectively record the use of lead as an additive in gasoline, and a world-wide increase in industrial mercury emissions. Thus, the lakes continue to compile and preserve records of all our activities – a record which future scientists may tease apart to better understand our culture and its relationship to current and future environmental problems.

Historical records portray recent climatic variability (Figure 7-6). The trends indicate that the Okanagan climate has warmed by about 1°C since 1900 AD. For early European settlers it was not unusual to cross Okanagan Lake on winter ice, an impossible feat in recent years. Factors contributing to the temperature variations and warming trend likely include natural changes in ocean circulation (for example, El Nino), volcanic activity and solar output, in addition to land-use changes, urbanization and greenhouse gas emissions. On a global basis, one recent assessment attributed about half of the temperature increase over the past 140 years to greenhouse gases. A natural increase in solar activity explained much of the remaining increase. Regardless of what is causing global warming, local paleoenvironmental studies of former warmer climates represent the keys to future planning. At stake are supplying towns with water, the trees that will optimize forest production, preservation of recreational fishing, and the prospects for the fruit and viticultural industries, to name a few.

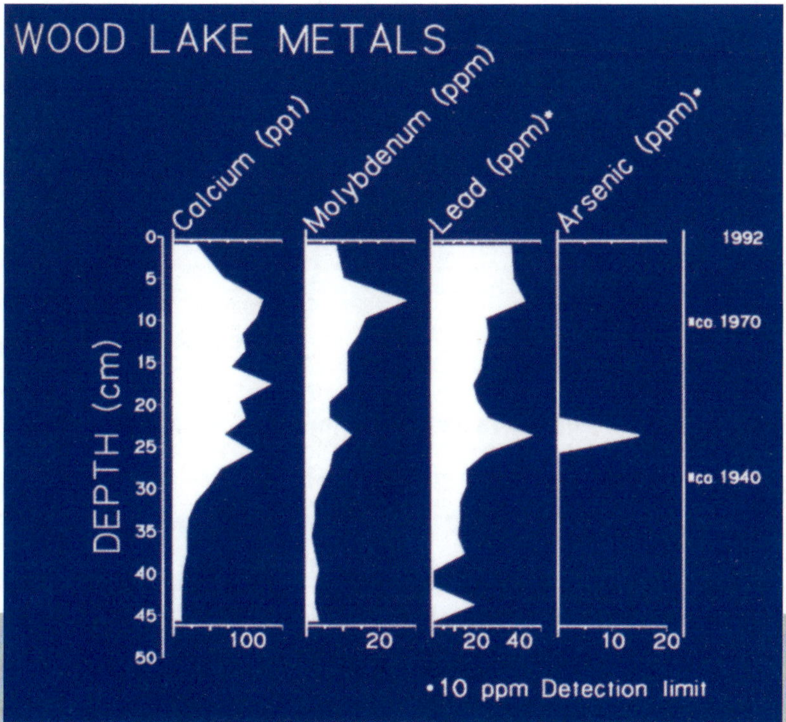

Figure 6-6. Stratigraphy of selected metals in the uppermost sediments of Wood Lake (Adapted from Walker et al., 1993).

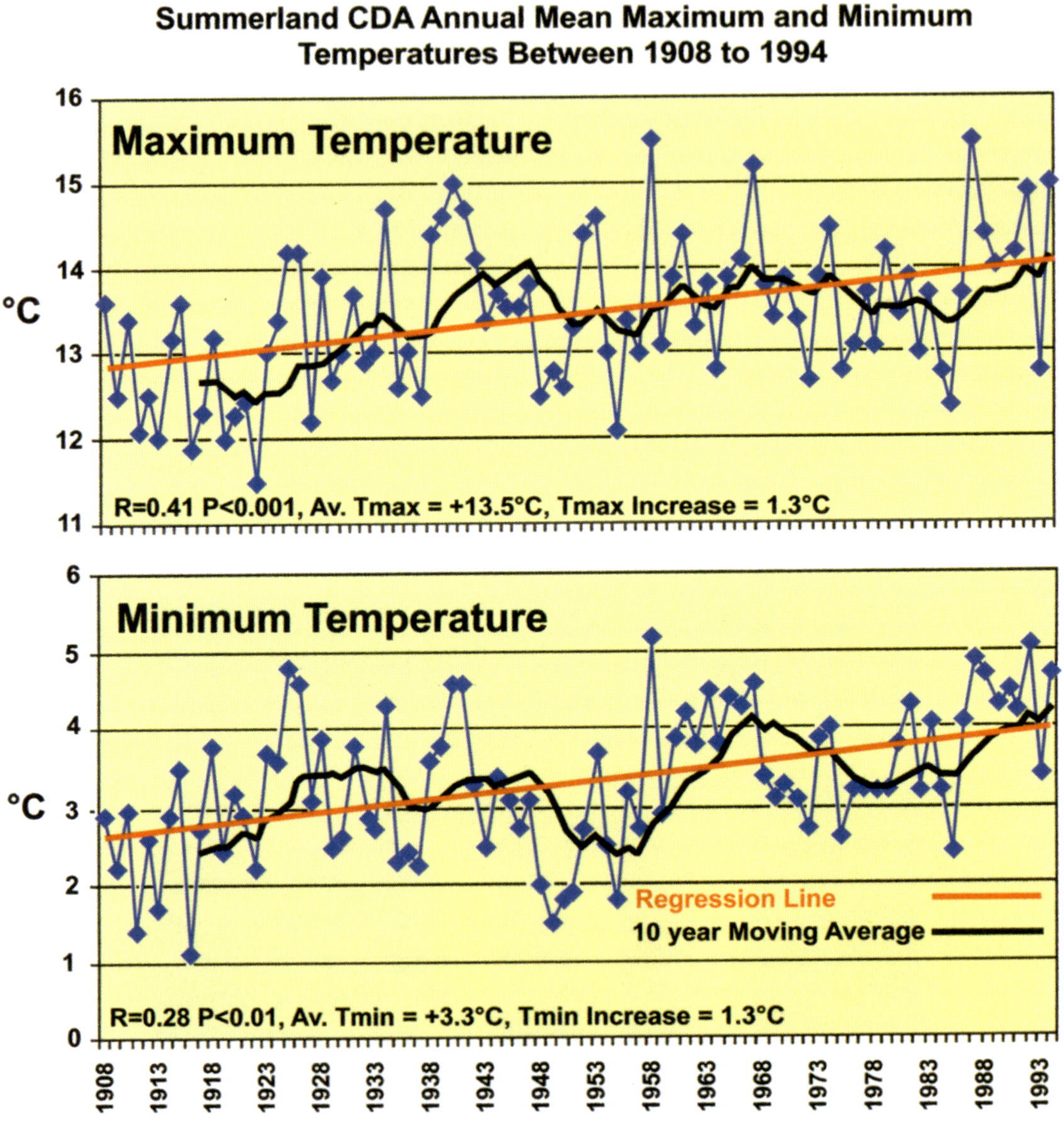

Figure 7-6. Historical temperature record from the Pacific Agrifood Research Station in Summerland. (Adapted from: Environment Canada http://www.ecoinfo.ec.gc.ca/env_ind/region/climate/climate_data_e.cfm)

SOIL LANDSCAPES & AGRICULTURE OF THE OKANAGAN VALLEY

by Herb A. Luttmerding

Soil in the pedological sense refers to that part of the earth's surface that supports plant growth. It consists of a zone at the surface usually less than a metre thick. Soil is formed by the interaction of soil parent material (surficial geologic deposits), climate, vegetation and landscape position (moisture conditions), all integrated over time. In the Okanagan Valley, all soil parent materials are of glacial or post-glacial origin and age. Thus, the last ten thousand years is an important geologic time interval for soil formation and has resulted in the development of a broad range of soil landscapes.

The general distribution and characteristics of Okanagan soils are given in Figure 8-6. In the south Okanagan area, between Penticton and the U.S.A. border, the valley bottom soils are mainly developed in sandy and gravelly, coarse textured glaciofluvial and alluvial deposits that are well or rapidly drained and developed under grassland vegetation. These are mainly Orthic Dark Brown or Orthic Brown Chernozem soils. Where topography is suitable, these soils support heat-loving tree fruits, vegetables and grapes, all produced under irrigation. The alluvial floodplain of the Okanagan River has sandy to silty, imperfectly to poorly drained soils, often strongly calcareous; they have Regosolic or Gleysoic soil development.

South Okanagan valley slopes mainly have moderately coarse textured, well drained soils developed in morainal or colluvial materials, commonly interspersed with bedrock outcrops. Soil development ranges from Orthic Dark Brown or Orthic Black Chernozems in grassland areas to Eutric Brunisols under forest. Land use is mainly cattle grazing and forest production. The valley slopes are also important wildlife areas, particularly during the winter.

Further north, between Penticton and Summerland, coarse textured soils of glaciofluvial and fluvial origin with Orthic Dark Brown or Eutric Brunisol soil development are prevalent. In the Summerland area particularly, fine sandy to silty eolian cappings overlie the gravels and sands. Another notable feature of the Summerland - Penticton - Naramata area is the silty glaciolacustrine terraces (Glacial Lake Penticton) flanking Okanagan Lake. The soils on these deposits are mostly Brown Chernozems, strongly gullied and susceptible to piping (Figure 7-10). The subsoils are also strongly calcareous. Irrigated tree fruits are the usual crops in this area.

The valley widens to several kilometres in the Westbank - Kelowna - Winfield area. Coarse textured, rapidly drained soils are still common although other soils prevail as well. Silty to clayey, well or moderately well drained soils of glaciolacustrine origin are common in the Westbank area and in the Glenmore Valley. Soil development here is dominantly Gray Luvisol with some Orthic

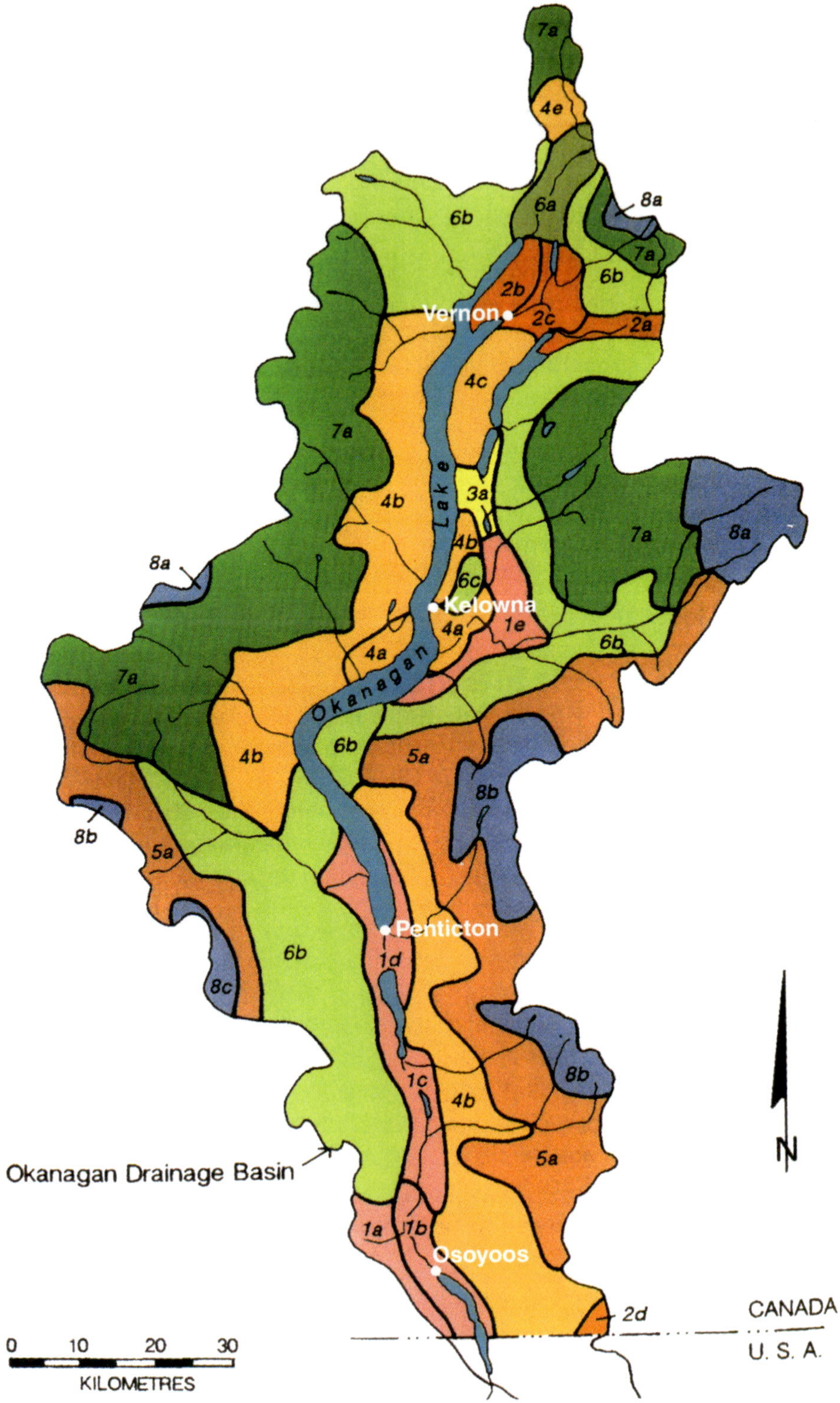

Figure 8-6 (opposite page): Soil landscapes of the Okanagan watershed (modified after Agriculture Canada, 1993).

LEGEND

SOIL LANDSCAPES OF THE OKANAGAN WATERSHED

1. DOMINANTLY DARK BROWN CHERNOZEM SOILS
   - 1A Morainal, inclusions of Brown soils and Eutric Brunisol colluvial soils.
   - 1B Glaciofluvial and alluvial, inclusions of Brown Soils.
   - 1C Glaciofluvial, inclusions of Regosolic alluvium.
   - 1D Glaciofluvial, substantial inclusions of silty Brown lacustrine soils.
   - 1E Glaciofluvial and morainal soils.
2. DOMINANTLY BLACK CHERNOZEM SOILS
   - 2A Alluvial, inclusions of Gleysolic soils.
   - 2B Morainal, inclusions of lithic colluvial soils.
   - 2C Morainal, substantial inclusions of clayey Dark Gray lacustrine soils.
   - 2D Loamy morainal, inclusions of lithic Eutric Brunisol colluvial soils.
3. DOMINANTLY DARK GRAY CHERNOZEM SOILS
   - 3A Glaciofluvial, inclusions of moderately coarse Dark Brown morainal soils.
4. DOMINANTLY EUTRIC BRUNISOL SOILS
   - 4A Morainal, inclusions of clayey Gray Luvisol lacustrine soils.
   - 4B Colluvial, inclusions of loamy Gray Luvisol morainal soils and rock outcrop.
   - 4C Colluvial, inclusions of loamy Black morainal soils and rock outcrop.
   - 4D Lithic colluvial, substantial inclusions of rock outcrop.
   - 4E Morainal, inclusions of coarse Black glaciofluvial soils.
5. DOMINANTLY DYSTRIC BRUNISOL SOILS
   - 5A Morainal, inclusions of coarse lithic colluvial soils.
6. DOMINANTLY GRAY LUVISOL SOILS
   - 6A Lacustrine, clayey, inclusions of coarse glaciofluvial Eutric Brunisol soils.
   - 6B Morainal, inclusions of moderately coarse lithic Eutric Brunisols soils.
   - 6C Lacustrine, clayey, inclusions of Dark Brown and moderately coarse Eutric Brunisol colluvial soils.
7. DOMINANTLY BRUNISOLIC GRAY LUVISOL SOILS
   - 7A Morainal, inclusions of moderately coarse lithic Dystric Brunisol colluvial soils.
8. DOMINANTLY HUMO-FERRIC PODZOL SOILS
   - 8A Morainal, inclusions of lithic colluvial soils.
   - 8B Morainal, inclusions of lithic colluvial soils.
   - 8C Lithic colluvial, inclusions of loamy morainal soils.
9. DOMINANTLY GLEYSOLIC SOILS
   - 9A Alluvial, poorly drained, substantial inclusions of coarse loamy imperfectly drained Regosolic soils.

Dark Brown Chernozems. Moderately textured, Eutric Brunisol or Gray Luvisol soils of morainal origin are also common, particularly along the valley sides. The fluvial delta/fan of Mission Creek occupies a large area on the east side of Okanagan Lake in the vicinity of Kelowna. A water table is usually present at various depths and soil development of these sandy to silty, moderately to poorly drained soils ranges from Gleyed Regosols to various Gleysols. A few small organic areas are also present. Agricultural use of the soils in the Westbank - Kelowna - Winfield area is mainly for tree fruits and grapes although some forage crops and vegetables are also produced. Irrigation is usual. The area is also subject to strong urbanization pressures.

In the Vernon - Armstrong area, much of the valley bottom is dominated by soils developed in clayey glaciolacustrine sediments. Well drained, moderately pervious Gray Luvisol soils (also some Solonetzic soils) are common and mostly used for non-irrigated forage and small grain production. Sandy and gravelly fluvial and glaciofluvial soils predominate on Grandview Flats and in parts of the Coldstream Valley. These well and rapidly drained soils have thick surface layers enriched with organic matter, classified as Orthic Black Chernozems. Agricultural production on these soils, essentially all irrigated, ranges from forage and silage corn to tree fruits and small fruits and vegetables. Most of the higher locations in the valley consist of medium or moderately coarse textured soils developed in morainal material. These soils are well drained sometimes stony Gray Luvisols or Eutric Brunisols.

The Okanagan Valley is deeply incised into the Thompson Plateau by at least 1,000 metres. Slopes on the flank of this upland north from Penticton are mainly forested except for some south facing parts of tributary valleys that are grass and shrub covered. The soils are moderately coarse textured, well drained and steep or moderately steep. They are developed in thin morainal material and colluvium, either over bedrock or interspersed with bedrock outcrops. Soil development consists dominantly of Eutric Brunisols with Black Chernozems in grassy areas.

The rolling forested plateau or upland flanking the Okanagan Valley is mantled with deposits of glacial moraine origin. The soils are generally moderately coarse textured, well or moderately well drained but include scattered, depressional, poorly drained organic areas. The usual soil development ranges from neutral Eutric Brunisols at lower elevations through Dystric Brunisols to strongly acid Podzols at the highest elevations. Gray Luvisols also occur where the soil parent material is medium textured.

Climatically, the Okanagan Valley is driest and hottest in the south and gradually becomes somewhat cooler and moister in the north. Similar conditions also occur with increasing elevation from the valley floor to the upland region. Soil development reflects this climatic variability, with Dark Brown and Brown Chernozems dominating in the south and Dystric Brunisols or Podzols the usual development at upper elevations.

CHAPTER 7

GEOLOGY AND TERROIR OF OKANAGAN VALLEY WINE

by John Greenough, Herb Lutmerding and Leanne Mallory-Greenough

---

The Okanagan Valley has earned an international reputation for world-class wines. Climate and soils play major roles in producing fine wines, and the N49 to N50 latitude of our grape-growing region is comparable to renowned wine-producing areas such as Champagne and Alsace in France. The suitability for viticulture is enhanced by a rain shadow from the Coastal Mountains which leads to hot dry summers. Similarly, large lakes moderate winter climate allowing delicate vitis vinifera vines to thrive. Thus, the influence of geologic parent materials on wines of the Okanagan has become an interesting avenue of research for geological specialists.

THE FIRST GRAPES

It is not known who first planted European vitis vinifera grapes in the Okanagan but the Summerland research station had experiments in the early 1900's and the Rittich family had trial vineyards in the 1930's and 1940's (Davis et al., 1984; J. Vielvoye, personal communication, 2004). Joe Busnardo made successful commercial plantings of vitis vinifera in the 1960's, but labrusca and hybrid-type grapes dominated the industry until 1988. Today, vinifera varietals, in order of production include Merlot, Chardonnay, Pinot Noir, Pinot Gris, Pinot Blanc, Cabernet Sauvignon, Gewurztraminer, Cabernet Franc and Riesling. Most have won major international awards (Fulton, 2003). The valley is famous for its ice wines, commonly considered the best in the world.

COMPARATIVE SIZE OF THE OKANAGAN VINEYARDS

Estimates of vineyard acreages vary but, for perspective, there are only about 4,200 acres in the Okanagan compared to 14,000 in Niagara, Ontario and perhaps 565,000 in the Napa Valley of California (Fulton, 2003). Nevertheless, the products are diverse, perhaps because of geological and climatological variations. There are changes in temperature from north to south, correlated soil differences, and variable bedrock and glacial sediment types. In addition, local microclimates reflect air-funneling mountain passes, differences in elevation, slope angle and direction, and proximity to lakes. The French long ago connected wine quality with bedrock, soil type and local climate that together describe the terroir of a grape-growing area. Variables contributing to the terroir of Okanagan wines are explored below.

TYPES OF BEDROCK, SEDIMENTS AND SOILS

Many different bedrock types underlie Okanagan vineyards. The eastern side of the valley is dominated by "granitic gneiss" of the Precambrian Monashee complex but locally there are many other metamorphic rock types (e.g. greenstone). Rocks on the western side of the valley include younger (Paleozoic to Mesozoic) metamorphic and sedimentary rocks (sandstone, limestone) intruded by Jurassic granites and overlain by Tertiary volcanic rocks and sediments.

Attempts to relate wine quality and composition to bedrock geology have been largely unsuccessful (Taylor et al., 2002) possibly because many vineyard soils occur on thick (10's to 100's of metres) accumulations of glacial to postglacial sediments (Fulton, 2003). The Fraser ice sheet (10,000 to 20,000 years ago) left behind till deposits. As it melted, glacial lake sediments (sand, silt and clay) accumulated along the margins of Glacial Glacial Lake Penticton, with sand and gravel deposits forming where meltwater rivers entered the lake. The sediments have been locally reworked into alluvial fan, delta and floodplain deposits. These young, highly permeable (water passes through easily) materials are perched along valley walls, high above the water table, and are exceptionally well drained. This is important. Grape vines flourish in well-drained soils.

OKANAGAN WINE REGIONS

The grape-producing area can be subdivided into three regions (Fulton, 2003): Okanagan River Region (U.S. border to Skaha Lake); Southern Okanagan Lake Region, (Skaha Lake to Peachland); Central Okanagan Lake Region (Peachland to Winfield). Vineyards in the Okanagan River region (Figure 1-7) are planted on well-drained, river-formed flood plains and alluvial fans. Floodplain deposits in the valley bottom host few vineyards because they are frost-prone and close to the water table. Benches of fine-grained (sand and silt) glacial lake sediment dominate the Southern Okanagan Region, the most important being the Naramata Bench (Figure 2-7). Vineyard soils of the Central Okanagan Lake Region formed from mixtures of glacial lake silt and clay, glacial river deposits and till (Figure 3-7). Although less permeable, thin sand and gravel layers at the surface, plus gentle slopes toward the centre of the valley, create good drainage.

The bulk chemical compositions of these sediments are all very similar because they were originally derived from a vast area, mixed, and deposited in the valley by glaciers (Greenough et al., 1996). However, climate has created variations in the soil types (Fulton, 2003). Hot and dry conditions in the Southern Okanagan produced Brown Chernozem soils. Cooler and moister conditions near Kelowna have yielded Dark Brown Chernozems or, in forested areas, Eluviated Eutric Brunisols.

CLIMATE, WINE COMPOSITION AND WINE QUALITY

Expert wine tasters can identify the country, region and even vineyard-of-origin of wines.

Obviously "terroir" contributes to wine quality and character. One parametre, Growing Degree Days, measures the total amount of useful heat received by vines during the growing season (Davis et al., 1984). Values range from 1500 degree days in the southern Okanagan to 1100 in the north and demonstrate a significant climate gradient in the valley. Valley side (east versus west), orientation, elevation and slope of vineyards also affect growing degree days. If these "terroir" parametres and their impact on wines can be quantified, the information might be used to optimize the grape varieties grown or certify the place of origin of a wine.

RESEARCH RESULTS

Research has related wine composition with the geographic origin of Okanagan grapes. The concentrations of elements such as Potassium, Strontium, and Nickel in wines show that individual vineyards have a "signature" (Greenough et al., 1997; Taylor et al., 2002). Concentrations also correlate with latitude and geographic location in the valley indicating that climate influences what is taken up by vine roots (Greenough et al., 1996). Niagara (Ontario) wines have an entirely different element composition (Taylor et al., 2003) suggesting that wines can be regionally "fingerprinted" and perhaps used to certify origin. However, element concentrations do not necessarily correlate with wine quality or character (bouquet and taste) that are determined by organic compounds. A study of compounds, including alcohols (e.g. ethanol = CH3-CH2-OH), acids (e.g. acetic acid = vinegar), and esters (fruit-flavored compounds) found that wines can be distinguished based on whether their grapes came from Oliver, Naramata or Kelowna (Eggers et al., 2004). Apparently taste also reflects climate variations and geography. Staff at the Pacific Agrifood Research Centre (Pat Bowen, personal communication, 2004) have defined Okanagan grape-growing regions based on climate, soils and terrain and are using grape chemistry, and ratings of fruit and wine quality, to optimize the production of high-quality wines. Clearly, the importance of terroir is appreciated and will enhance the Okanagan's reputation as a great wine producing area.

Figure 1-7. A photograph of vineyards in the south Okanagan showing sparsely- treed valley walls resulting from a dry, semi-arid climate. John D. Greenough, photo

Figure 2-7. A vineyard in the Naramata area looking across the lake to Summerland. The vineyard is planted on a "bench" underlain by fine-grained, Glacial Lake Penticton sediments. Giants Head in Summerland is at the far left.
John D. Greenough, photo

Figure 3-7. Scenery showing vineyards planted on sloping benches in the Mission area of Kelowna. Coniferous trees, apparent in the distance attest to a damper, more temperate climate than in the southern Okanagan. Lighting conditions and haze around the distant hills of Okanagan Mountain Park are products of the summer 2003 Kelowna fire which had started in the park 36 hours earlier.
John D. Greenough, photo

CHAPTER 8

ANCIENT PEOPLES OF THE OKANAGAN

by George Ewonus, Paul Ewonus and James Baker

Controversy still exists about the dates and routes of the general peopling of North America, much less the more specific questions of aboriginal settlement in the Okanagan. Reliable archeological evidence supports people settling in the Americas at least 15,000 years ago. Many of the sites that support this date are south of the Okanagan for the obvious reason that the land here was covered by ice as recently as 15,000 years ago. However, there were people living south of the Okanagan in the Columbia River basin about 11,000 years ago. We also know that there were people living in northern British Columbia at the Charlie Lake Cave site at least 10,700 years ago (Fladmark, 1988). It is presumed that it was the northeastward migration of these people, up the Columbia River basin, and possibly the Fraser River basin, that led to the peopling of the Okanagan some 9,000 years ago (Borden, 1979; Fladmark, 1983; Baker, 1990). So far exact dates of first settlement are unknown because very little systematic archeological work has been done in the Okanagan valley area. It is plausible however that the higher elevations of the Okanagan Upland were visited first since there may have been a stillstand during the retreat of the Fraser Glacier about 13,000 years ago (see Chapter 5) by which time ice above 900 metres had melted.

THE OKANAGANS

While there has been very little study of the earliest peoples of the Okanagan, we do know a little more about their likely descendants, the Okanagan people of 3,000 to 5,000 years ago. At the height of Okanagan culture approximately 3,000 years ago it is estimated some 12,000 people lived in this valley and the surrounding areas (Baker, 1990). Linguistic and ethnographic data suggest that traditional Okanagan territory included the Okanagan Valley as well as the Arrow and Slocan Lakes (Figure 1-8). In addition, areas south of the 49th parallel included the Okanagan River south to Brewster as well as Kettle Falls and the Colville River (Turner, Bouchard and Kennedy, 1980; Baker, 1990). Within this approximately 72,000 square kilometres of territory some seven dialects of the Okanagan language were spoken (Turner et al., 1980). The Okanagan language is part of the Interior Salish language division of the Salishan language family. The Okanagan people share numerous cultural features that are common to groups of the Plateau culture area.

OKANAGAN PLATEAU PITHOUSE TRADITION

Most of the limited archeological research about the Okanagans has concerned the period from 4,500 to 200 years before present (BP). The only regional synthesis developed for the Fraser Plateau based upon rigorous comparative analysis is that of Richards and Rousseau (1987; Stryd and Rousseau 1996) for the late prehistoric period. On the basis of habitations, economic adaptations, and stone tool (lithic) assemblages, they defined the Plateau Pithouse Tradition (PPT), spanning the last 4500 years BP. During the PPT, lifeways were considered to be generally similar to those described for the ethnographic Interior Salish of the Canadian Plateau. The period can be characterized by the use of semi-subterranean pithouses, commonly known as "kekulis," as dwellings in semi-permanent winter villages. People employed a hunter-gatherer adaptive strategy with an emphasis on salmon fishing and food storage in underground pits. The pattern of settlement was semi-permanent, and resource acquisition was logistically organized and seasonally regulated. The relative importance of fishing, hunting and gathering varied over time and between regions, although the diet throughout was primarily composed of anadromous salmon, supplemented by deer, elk, a variety of small mammals, non-anadromous fish, fresh water mussels, birds and an assortment of plant foods such as roots and berries.

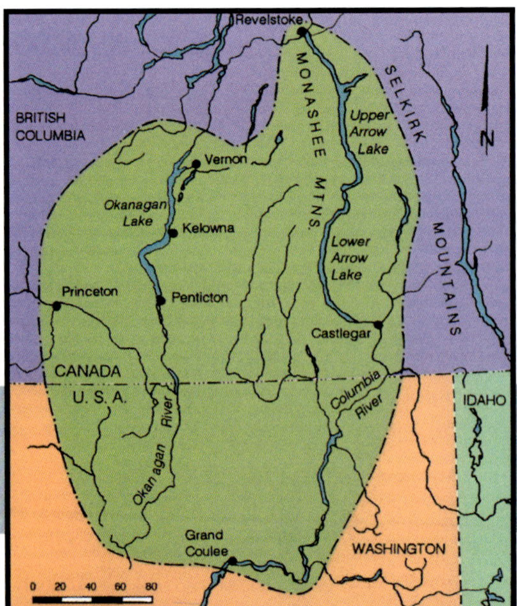

Figure 1-8: Traditional Okanagan Territory.

OKANAGAN WAY OF LIFE

The Okanagan people of the ethnographic period lived a rich life moving within traditional use areas throughout the year to fish, hunt or collect food, depending on the season. In this sense they may be called hunter-gatherers. In the Winter months they lived in villages of energy efficient pithouses accommodating 6 to 10 extended family members. Pithouse construction involved digging out a large circular or rectanguloid area of earth to a depth of approximately one metre. Four large posts served as structural roof supports. The roof was then covered with a lattice of smaller branches and earth. These structures were well insulated, resulting in an exceptionally warm and heat efficient dwelling throughout the cold winter months. There was a squarish opening at the top created by the four main roof rafters, functioning as a smoke-hole and at times a defensible entrance. Occasionally there was a side entrance to the pithouse, either in place of or in addition to the rooftop opening. Villages might contain anywhere from four to over one hundred such dwellings. Each village was largely autonomous with its own headman or chief, depending on size (Hudson, 1990). Numerous pithouse village settlements have been destroyed by development in the Okanagan valley. Nevertheless, protected areas such as Okanagan Mountain Park still contain relatively undisturbed sites. While large collections of pithouses have not been examined in the Okanagan, in the Kamloops area a collection of more than 200 pithouses has been recorded (Wilson and Carlson, 1980) and several large pithouse sites exist near Lillooct. Over time there were changes in pithouse design, possibly reflecting climatic and social changes. In addition to the pithouses, there were also smaller cache pits used to store food, as well as pits to roast plant and amimal foods. An example of a pithouse is on display at the Kelowna Museum. Full scale reconstructions can be found at the Secwepemc Cultural Education Center in Kamloops and the Xatsu'll Heritage Village at Soda Creek, north of Williams Lake and also at Quaaout Lodge near Squilax.

SOCIAL ORGANIZATION AND ACTIVITIES

The Okanagan way of life involved many group activities while maintaining a high degree of family and personal autonomy. Winter villages could be occupied by a few or a great many families. Main winter villages, with many families, would have a single chief. Each large winter village would be considered a band. Membership within a village was fluid and families were free to move from one village to another (Hudson, 1990; Bailey and Rousseau, 1994). Positions such as the Salmon Chief and hunt leader were chosen on ability and spiritual powers. Interaction between villages often occurred at communal food gathering areas, especially during fishing season. There were also other social occasions that promoted festive gathering. On June 30th, 1877, the geologist George Dawson while on a trip along the old aboriginal trail near Naramata noted a large group of Okanagans on their way home from a "great potlatch" held at the head of the lake [Vernon area] (Cole and Lockner, 1989). This socialization was likely a longstanding custom. It provided an opportunity for the Okanagan people living at the foot of the lake [now Penticton] to interact and trade with other Okanagans from the head of the lake.

FOOD GATHERING

In the Spring the extended family groups would move from the winter village to their spring sites which would involve a more mobile form of dwelling, possibly woven tule mat structures. These structures may have had a rectangular floor plan with an inverted "V" roof covered with woven tule mats (Rousseau, 1994; Turner et al., 1980). Since the horse was not introduced into the Okanagan until after 1730 (Keyser, 1992), dogs harnessed to travois would be used to help transport goods. Early spring plants such as bitterroot (Lewisia rediviva), "Spitlem" in Okanagan language, would be harvested and processed in large rock-lined roasting pits according to traditional methods (Turner et al., 1980). Some of these roasting pits remain today, scattered throughout the Okanagan, an example of which can be found in Okanagan Mountain

Park. In addition to bitterroot, numerous other plants would be harvested along with year round staples such as fish and game.

FISHING

Fish were a major food resource for the Okanagans. Although all varieties of fish were utilized - trout, kokanee, coarse fish, sockeye and chinook salmon - salmon was the most important species. As with all food gathering activities there was an associated pattern and ritual which served to both conserve and distribute the resource equitably. There was a "Salmon Chief" who not only directed the construction of the weirs and traps but also controlled the distribution of the catch (Hudson, 1990). Salmon fishing could be significant social occasions. Since all members of the family could be involved in the fishing and processing, this activity might stretch over some weeks in the early Fall. This would be a good opportunity for families to come together, and marriages might be arranged at such times (Hudson, 1990). The major salmon fishing site in Okanagan territory was Okanagan Falls. Other sites available to the Okanagans were Kettle Falls and areas in Shuswap territory. The Okanagans may have also traded for salmon from sites not in their traditional territory. Both Okanagan and Kettle Falls localities were greatly affected by the reservoir of Grand Coulee Dam, but these areas still represent the best prospects for additional discoveries.

HUNTING

There were usually four major hunting expeditions during the year - the mid winter hunt, the spring, the fall and the early winter hunts (Teit, 1930). Of course, this could vary depending on need. Hunters could be away as long as one to two months, and distance traveled could be large. In the Fall, the hunters would travel to the Okanagan Highlands in search of game. At least one Okanagan Highland site has been recorded (Baker, 1975).

Large game included bear, elk (possibly Roosevelt), deer (both white tail and mule), moose, mountain goat and sheep. Smaller game included rabbit, beaver and grouse, duck, martin, turtle, squirrel, marmot and others. As with fishing there was traditionally a hunt leader as well as associated ritual both before and after all major hunts, which were group activities (Baker, 1990). The Fall hunt would concentrate on bear, elk, moose and deer while the late winter was mainly for sheep. In mid-winter the hunt would concentrate on deer, especially if the snow was crusted enough to support the hunters, but not the deer. Smaller game were hunted all year and formed an important part of subsistence. There are several known pictograph panels that record hunting activities (Corner, 1968).

OKANAGAN VALLEY ARCHAEOLOGICAL SEQUENCE

Following what were likely similar beginnings 10 to 12,000 years ago, the northern and southern cultures of the Okanagan Valley developed along somewhat different lines. The north became similar to the Thompson/ Shuswap region, while the south came to resemble the middle Columbia Plateau. The southern Okanagan culture historical sequence, as first outlined by Grabert (1974) and modified by Chatters (1986), consists of four phases: Okanagan (8000 to 6000 BP), Indian Dan (6000 to 3000 BP), Chiliwist (3000 to 900 BP) and Cassimer Bar (ca. 900 BP to AD 1850). Grabert first defined the characteristics of the Okanagan Phase on the basis of a number of small and apparently early assemblages from the valley, and acknowledged that this definition was provisional and rather poorly supported. Okanagan Phase sites were either stratigraphically early or located on the higher river and lake terraces, although no radiocarbon ages could be obtained. The Okanagan Phase people as employing a geographically fluid settlement strategy, with the nuclear family as the primary residential group for the majority of the year. They spent most of the year along local rivers, although they did travel into the highlands periodically. Stone material use and tool production emphasized expediency and no evidence for storage of food is yet known. Dwellings consisted of small (2.5 to 3.5 m diameter) pole-framed huts or tipis covered with skins, grass, or mats and occupied in both summer and winter. Subsistence was generally opportunistic, with a limited variety of species taken in differing proportions depending on local availability. The most important food animals were jackrabbit, deer, coyote, river mussels and a variety of fish.

The succeeding Indian Dan Phase is distinguished by large basal-notched stemmed spear points (some with distinct barbs), medium and large leaf-shaped points, pestles, milling stones, and numerous flake tools. Sites are commonly situated in rock shelters or open locations and earth ovens are sometimes present. Peoples of this time were more sedentary hunters and fishermen, living for years at a time in extended family groups housed in square to circular pithouses. Groups preferentially hunted deer, elk, sheep, and pronghorn antelope using darts armed with stone points, but pursued a substantial array of other animals when necessary. Also taken were various carnivores, rabbits, porcupine, beaver, marmot, suckers and salmon, mussels, and a variety of birds. Berries and undoubtedly many other plant foods were also harvested. Shortfalls in resources were ameliorated by a well developed trade network that brought in stone and marine shells for tools and ornaments. However, there is no available evidence for a developed storage technology during this time.

During the Chiliwist Phase activities were tightly scheduled to take advantage of spatial and seasonal concentrations of food resources. The most important animals were salmon, river mussels, and ungulates such as deer, elk, mountain sheep and pronghorn antelope. They were apparently harvested in quantity and preserved for consumption during the winter. Families occupied villages of deep, steep-walled, circular housepits and stone tool curation was generally emphasized over expediency. The tool kit included large leaf-shaped points, medium-sized barbed basal-notched stemmed points, microblades, milling stones, ground stone adzes, a variety of knives, scrapers, gravers, and occasional bone artifacts, including valves for composite harpoon points. A wide variety of raw material types are evidenced in the stone tool industry, and exotic shells and obsidian indicate the existence of extra-regional trade. Copp (1979), on the basis of work at the McCall site (DhQv 48) near Okanagan Falls, has proposed three subphases of the Chiliwist Phase. The last subphase (Subphase III, 1800 - 900 BP) is a 900 year hiatus in the archaeological record of the region. Subphase I (3000 - 2400 BP) is distinguished primarily by the presence of microblades, and Subphase II (2400 - 1800 BP) by a series of projectile points similar to, but smaller than, those of the first subphase. Furthermore, Copp interprets

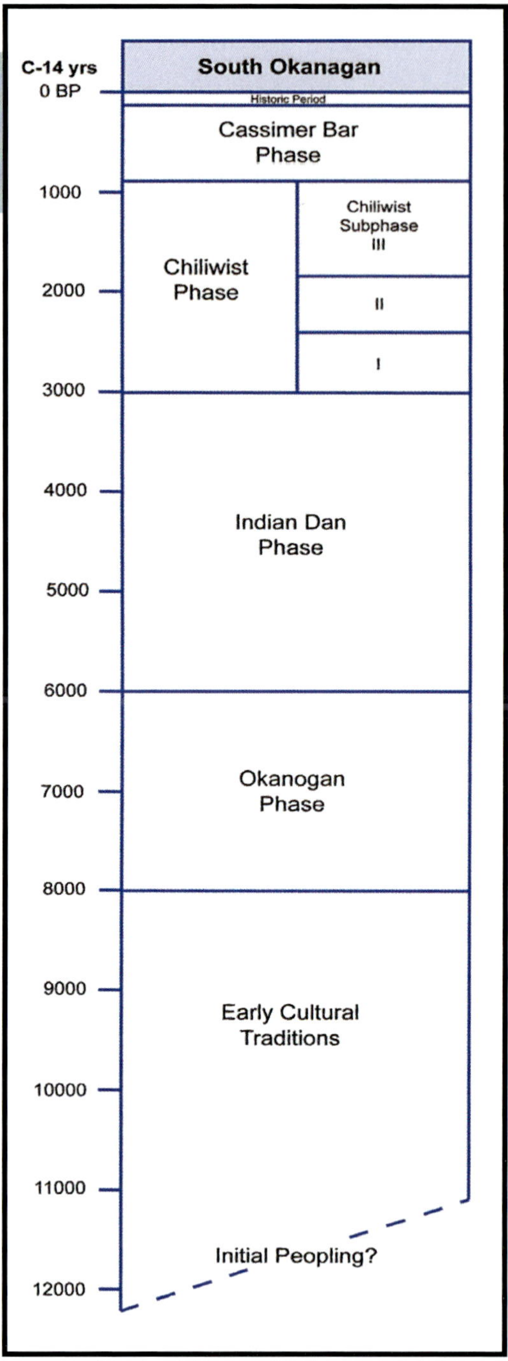

Figure 2-8. South Okanagan archeological sequence, based on Chatters 1986 and Copp 1979.

the shift in dominance from wide-necked to narrow-necked points late in Subphase II as indicative of the introduction of bow and arrow technology and the partial displacement of the atlatl, or throwing spear.

During the following Cassimer Bar Phase, circular or ovoid, shallow, saucer-shaped housepits and rectangular matlodge depressions are commonly found, as well as the reported occurrence of wood cyst burials. Mobility was greater than the previous period and social groups varied in size. Subsistence was focused on fish, ungulates, freshwater mussels and probably plant foods. Storage was well developed and tool manufacture continued to emphasize curation, with much the same tool kit as was in use late in the Chiliwist Phase. Formed artifacts include a variety of small arrow points, including corner-notched and side-notched forms, carved steatite items, milling stones, and bone composite toggling harpoon heads. Several bone and stone artifacts are incised with geometric or zoomorphic designs. In general, the way of life evident during the Cassimer Bar Phase was much like that described ethnographically for the southern Okanagan (Bicchieri, 1990; Chatters, 1986; Grabert, 1974; Pokotylo and Mitchell, 1998).

TRADITIONAL OKANAGAN LITHIC RESOURCES

Most of the named phases in the above assemblages are defined by the changes through time that occur in the stone tools used in the traditional life-way of the Okanagan people. A variety of fine grained, easily worked, stone material was used for the manufacture of the necessary implements. The various lithic materials included obsidian, chert, chalcedony, rhyolite, quartzite, quartz crystal, petrified wood and dacite. The Okanagan has a number of locations where fine grained stone material was collected for the purpose of stone tool manufacture. No systematic, extensive, intensive archaeological survey has been done to locate all Okanagan quarry sites.

Jim Pike (1975) conducted a small scale survey for quarry sites in parts of British Columbia. He recorded a number of quarry sites of materials suitable for stone tool manufacture in most parts of the province. Some of the sites in Okanagan territory or in close proximity to the Okanagan include rhyolite from Mt. Estakwalen and the Uppenborn site near the Greystokes; chert and chalcedonies from Sheep, Shatford and Shingle Creeks (W. Armstrong, pers. com. 1975); chert from Black Mountain (R. Robbins, pers. com. 1981); red jasper from Boulou Lake (R. Bonneau pers. com 1975); chert from Fir Valley; dacite from the Pennask area and Smith Creek (Mallory-Greenough, et al., 2002).

Differing amounts of the various materials were used at different times and at different locations throughout Okanagan territory. The one constant in the stone tool sequence is the use of dacite. Almost every excavated archaeological site that contains any quantity of stone tools has at least some made from dacite. This material under macroscopic examination "..reveals an extremely fine-grained and partially-glassy volcanic rock exhibiting low percentages of small phenocrysts."(Mallory-Greenough et al., 2002). In the archaeological literature dacite has been called basalt, glassy basalt, vitreous basalt, granular basalt, flint, and chert. Magne (1985) in his replication experiments of stone tool technologies in British Columbia says: ...The most common material of the Interior Plateau lithic technologies and in the experiments is high quality basalt that ranges in texture from vitreous to granular, and that is available as glacial till cobbles or stream bed cobbles in many areas (Magne, 1985:98). Geochemically speaking, it is not basalt but dacite Magne is referring to, and it is the most commonly used material in the Plateau Culture Area. Therefore it lends itself, when properly identified, as an excellent material through which to discuss trade patterns or travel routes in Okanagan territory.

Recent geochemical analysis (Commisso, 1997; Mallory-Greenough et al., 2002) has shown that dacite from different locations has a unique 'fingerprint.' This would enable archaeologists to identify the source of dacite recovered in excavation and comment on trade patterns or geographic mobility of the people of the Plateau Culture Area. Some of the more exotic lithics, clear quartz crystal, obsidian and petrified wood, of themselves imply trade or long distance travel. Their occurrence on the Plateau is rare, as is their recovery in excavation. Dacite is recovered in considerable quantity in early, middle and late archaeological assemblages from the Plateau but not all dacite sources provide the most desirable flaking characteristics for stone tool technology.

Five dacite sources in the south central interior of British Columbia, one near Cache Creek, another near the Similkameen River by Smith Creek, and three near Kelowna at Layer Cake and Boucherie Mountains and in the Pennask Lake area, were sampled and identified by geochemical analyis (Fig. 8 & 9 fr. Mallory-Greenough et al., 2002). No quarry activity was evident at the Layer Cake or Boucherie sources and the dacite from these locations is more difficult to work. Macroflakes or spall tools suitable for hide or bark scrapers could be formed with minimal retouch from this material, so it might occur in archaeological context. The dacite from the other three sources is easily worked and likely will be found throughout the archaeological sequence in the Okanagan.

A program to identify the source of dacite recovered in previous archaeological excavations in the central and south Okanagan is being developed. Chief Robert Louie and the Westbank First Nation Council have agreed to provide mate rial recovered in excavation on WFN lands and Stan Copp will provide material for analysis from some of his work in the south Okanagan. Results will be extremely interesting as identification will be from an important material that shows at least 9,000 years of use in stone tool technology on the Plateau. There will most likely be more sources to locate in Okanagan Territory and throughout the entire Plateau Culture Area in order to match material identified from archaeological collections.

PICTOGRAPHS, REFLECTIONS AND POTENTIAL

If as Baker (1990) suggests there were 12,000 people living in the larger Okanagan territory, then Okanagan life must have been rich and vibrant. Further evidence for this is recorded in the many pictographs throughout Okanagan territory. Pictograph themes represented vary from spiritual visions to hunting practices. While direct precision dating of the pictographs (Figure 5-8) has not yet been attempted, there is archeological evidence suggesting that some pictographs are at least 2,000 years old (Copp, 1980). A pictograph spall in the adjacent Columbia plateau has been archeologically dated to between 6,000 and 7,000 years ago (Keyser, 1992). Furthermore, a comprehensive inventory of pictographs in Okanagan territory has never been attempted, and although many have been discovered their locations are closely guarded secrets to protect against vandalism.

Often a pictograph panel may tell something about an adjacent village site. In the case of the Good Creek winter village, which may have also been a food processing site, a nearby pictograph panel depicts hunting activities (Bailey and Rousseau, 1994). Throughout the Okanagan valley many beaches and bays still show evidence of ancient Okanagan people's presence.

ROLE OF GEOLOGY

Earth Science has a direct role in assisting archeologists and this takes several forms.

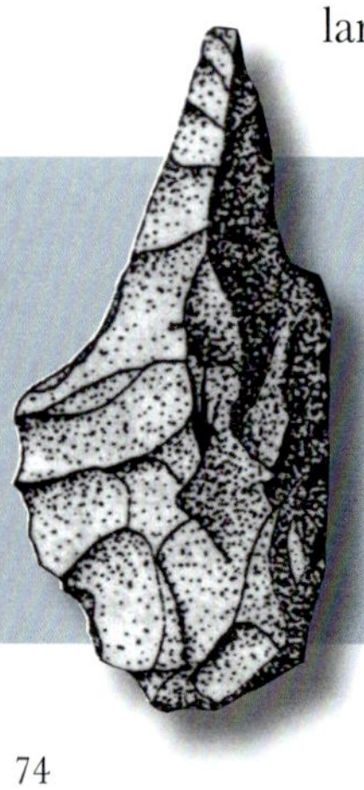

Figure 3-8. Large key-shaped formed uniface of medium grey chert from site DlQv-39, Westbank, BC (half size). This type of artifact was used primarily for perforating and scraping/carving tasks. They are known to date to the period between ca. 4000 and 1000 BP on the Columbia Plateau, and in the Okanagan specifically, to the Chiliwist phase (3000 to 900 BP; Rousseau 1992:28). Artifact number DlQv-39:27. Reproduced with permission of Antiquus Archaeological Consultants Ltd from Rousseau 1984:Figure 72b.

Firstly, terrain scientists are capable of identifying past terrestrial environments that may have controlled the early migration of the Okanagan people. The most obvious example is the stranded shoreline of Glacial Lake Penticton. This shoreline occurs at an elevation of between 500 and 525 metres and would have been a natural corridor to follow. In places there is a well developed sand beach in former embayments along the rock slopes of the valley. These are prime locations for early archeological sites (see Figure 19-9). Along the same line, geologists may be able to pin-point former streams that entered the lake, and these locations may have been particularly attractive to early people due to fishing potential. Even the location of warm or cold groundwater springs, especially in upland regions, may provide a starting point for future discoveries. Close observation of air photographs may reveal fracture zones and cliffs in the bedrock which could be ideal sites for pictographs.

When a site is discovered, details regarding the origin of the sediments deposited at the site as revealed by geologic study can be of great benefit in the interpretation of climatic conditions at the site.

Geologists and archeologists also complement each other's work by accumulating data that provide more precise chronologies of post-glacial historic events. This assists the archeologists in tracing the evolution of successive cultures and allows them to correlate to other more distant sequences of habitation. An example of this is the summary of the last ten thousand years provided in Chapter 6. It is now possible to include the archeological history in Figure 1-6, for example, which along with the other data begins to present a more comprehensive view of this time period.

Figure 5-8: A pictograph on a steep rock cliff in the South Okanagan displays man and animals. There are numerous pictograph sites in the Okanagan Valley region but no systematic and scientific inventories have been done.

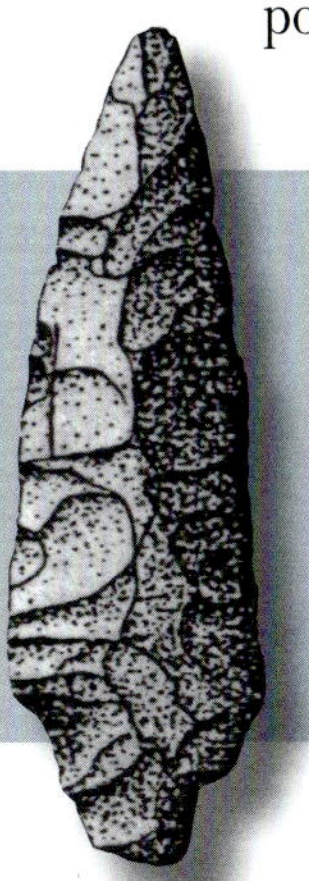

Figure 4-8. Medium sized, lanceolate, contracting stemmed projectile point of vitreous dacite from site DlQv-39, Westbank, BC (half size). This style of point dates approximately to the period between 4000 and 2000 BP. Artifact number DlQv-39:39. Reproduced with permission of Antiquus Archaeological Consultants Ltd from Rousseau 1984:Figure 71f.

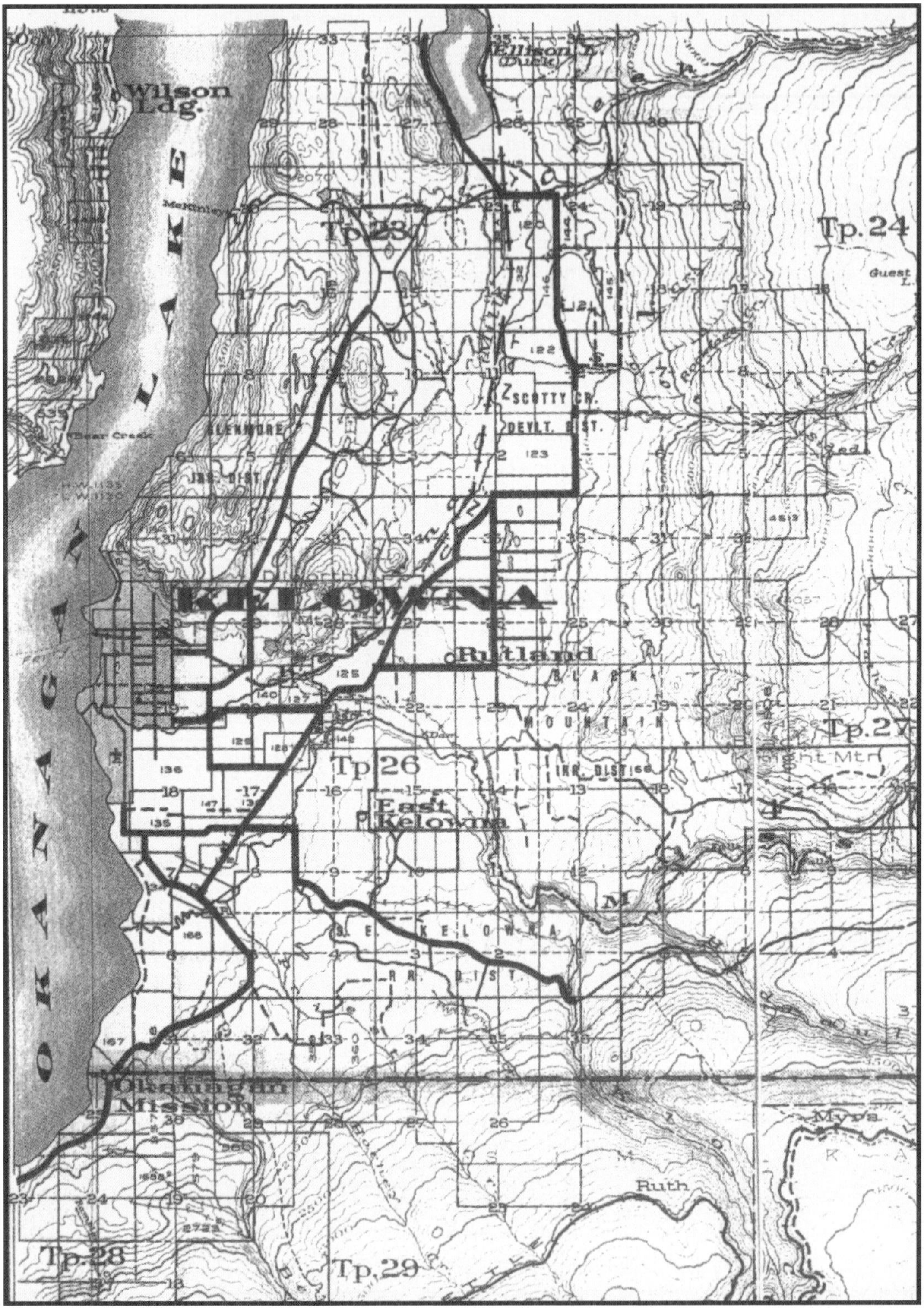

Figure 6-8: A portion of the "Kettle Valley Sheet", edition of January 31, 1923. This map was surveyed by J.D. McCaw for the Department of Lands, British Columbia. It is a topographic map at a scale of one inch = two miles.

FUTURE POTENTIAL

Approximately 350 archeological sites are now known in the vicinity of Okanagan Lake. Despite this, our understanding of the ancient peoples of the Okanagan is extremely limited. Collections of artifacts and systematic site descriptions are very rare. It is estimated that the Okanagan could have more than 1000 archeological sites, and there have been virtually no upland site discoveries. The current rate of development in the Okanagan Valley underscores the importance and need for further geological, ethnographic and archeological research, and offers students an exciting career alternative.

Figure 7-8: Acrylic painting by M. Roed. This spectacular hoodoo is located a few kilometres north of Falkland on the side of a hill overlooking Pillar Lake (used to be called Pink Lake). It is an erosional remnant of a conglomerate that, during Tertiary time, filled a portion of the Chase Creek Valley. Apparently this valley occupies a fault trench and received very coarse material from adjacent uplands. After several glaciations and extensive interglacial erosion, all that is left of this unique deposit is the "pillar" and similar rock outcrops that form steep slopes at the locality. Considerable mystery surrounds the precise origin of the pillar and the conglomerate from which it is composed. It has to be one of the most profound geologic landmarks in British Columbia.

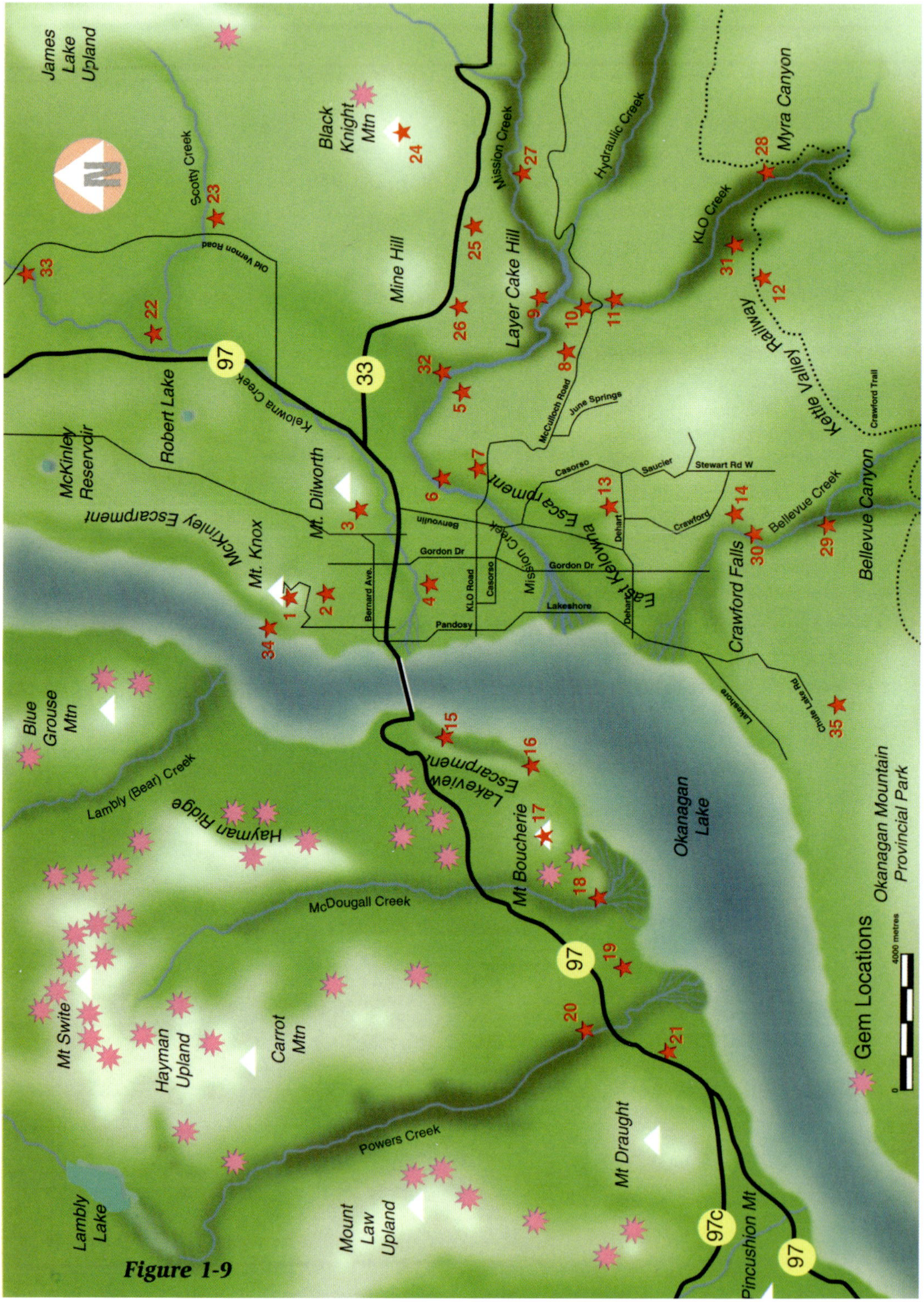

James Lake Upland
Black Knight Mtn
Mine Hill
Layer Cake Hill
Mission Creek
Hydraulic Creek
KLO Creek
Myra Canyon
Kettle Valley Railway
Crawford Trail
Scotty Creek
Old Vernon Road
Robert Lake
McKinley Reservoir
McKinley Escarpment
Kelowna Creek
Mt. Dilworth
Mt. Knox
Benvoulin
Gordon Dr
Bernard Ave.
KLO Road
Casorso
Pandosy
Lakeshore
McCulloch Road
June Springs
Saucier
Stewart Rd W
Dehart
Crawford
East Kelowna Escarpment
Crawford Falls
Bellevue Creek
Bellevue Canyon
Chute Lake Rd
Blue Grouse Mtn
Lambly (Bear) Creek
Hayman Ridge
Lakeview Escarpment
Mt Boucherie
McDougall Creek
Okanagan Lake
Okanagan Mountain Provincial Park
Gem Locations
4000 metres
Mt Swite
Hayman Upland
Carrot Mtn
Powers Creek
Mt Draught
Lambly Lake
Mount Law Upland
Pincushion Mt
Figure 1-9

Figure 1-9

CHAPTER 9

GEOLOGIC LANDMARKS OF THE KELOWNA AREA

by Murray A. Roed

Specific geological features that are local scenic landmarks in the Kelowna area are described in this Chapter which is intended to be used in conjunction with a self-guided tour called ***"A Sight-seer's Delight"*** in Appendix A. The features are numbered and shown along with main roads on the map of Figure 1-9.

LIST OF GEOLOGIC LANDMARKS IN FIGURE 1-9

1 Knox Mountain Viewpoint
2 Varved Silt Terraces, Glacial Lake Penticton
3 Mount Dilworth Escarpment
4 Flatlands of Kelowna and Rutland
5 Terraces of Rutland and the Rutland Fan
6 Mission Creek Regional Park
7 Varved Clay, Glacial Lake Penticton
8 Hummocky Moraine
9 Gallagher's Canyon
10 Till Outcrops
11 Quarry, Shuswap Rocks
12 Kettle Valley Railway, Plateau Basalt
13 East Kelowna Fan
14 Beach Deposit, Glacial Lake Penticton
15 Lambly Creek Basalt Cliff, Lakeview
16 Kalamoir Regional Park
17 Mount Boucherie Columns
18 Sandstone Cavities, McDougall Creek Fan
19 Varved Clay, Boucherie and Gellatly Roads
20 Glen Canyon Regional Park
21 Coal Beds, Fossil Trees
22 Kelowna Bog
23 Lateral Moraine
24 Roche Moutonnée
25 Eskers and Kames
26 Kettle Lake, Garner and Belgo Ponds
27 Mission Creek Waterfall
28 Myra Canyon
29 Bellevue Canyon
30 Crawford Falls
31 Warm Spring
32 Cutbank of Mission Creek, Rutland Fan
33 Mill (Kelowna) Creek Falls
34 Paul's Tomb
35 Kelowna Crags

Figure 1-9 (opposite page): Location map of Geologic Landmark Features in the Kelowna area. Bracketed numbers in text refer to numbers on the map, keyed to an accompanying list of the features. Note that a few of the features listed are not discussed in the text. Stars are gem localities.

KNOX MOUNTAIN VIEWPOINT (1)

From the observation station about a three minute walk from the parking lot at the top of Knox Mountain Park (Figure 1-9) many of the geologic features discussed in this and previous chapters can be observed in one spectacular panoramic view. Points of interest are shown in a directional rosette in Figure 2-9.

To the southeast, in the distance, the slanting rock slabs of the Precambrian Monashee Gneiss outcrop along the east side of Okanagan Lake. On the west side are hills in the Westbank area composed of volcanic and sedimentary rocks of the Tertiary Period (Eocene Epoch, 37 to 58 million years ago). Separating the two is the Okanagan Valley Fault zone which runs through East Kelowna and the Mission and then toward Vernon. One can appreciate the amount of vertical displacement on the Okanagan Valley Fault since the Tertiary time by noting that when the volcanic rocks were formed at the surface of the earth, the Monashee Gneiss rocks were buried several kilometres down in the earth!

The last glaciation had a dramatic effect in shaping the landforms visible from Knox Mountain. The ice was higher than the mountain tops visible in the distance! The bedrock base of Okanagan Lake lies almost one kilometre below your feet, acting as testament to the incredible grinding power of the ice sheet. Deep scratches (striae) etched into the rocks at the top of Knox Mountain (within 30 metres to the north of the lookout point) by boulders embedded in the base of the Fraser Glacier indicate the local direction of ice movement.

GLACIAL LAKE PENTICTON SILT TERRACES (2)

Nestled against Knox and Dilworth Mountains are sediment remnants of the extensive lake that developed as the Fraser Glacier melted. Glacial Lake Penticton reached an elevation of about 460 metres or 1500 feet above sea level which is over 100 metres (370 feet) above the present level of Okanagan Lake.

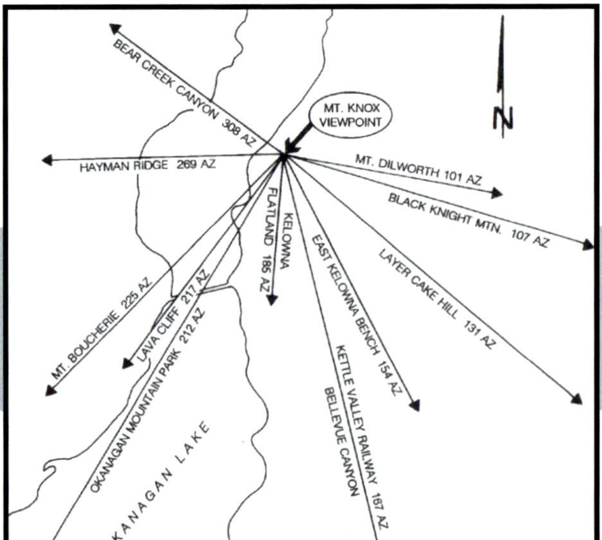

Figure 2-9: Directional rosette for the main Knox Mountain viewpoint. Arrows indicate the direction in degrees Azimuth. For example, 180 AZ would be due South.

Figure 3-9: A view to the south from Knox Mountain viewpoint shows dipping metamorphic strata in the distant Okanagan Mountain Park and a volcanic dome, Mount Boucherie, looming at the right. New posters have been installed in 2014.

Deposits formed in the glacial lake dominate the lower valley slopes forming bench-like landforms on both sides of Okanagan Valley and above the flatland of the main part of the City (Figures 3-9 and 4-9). The deposits consist mainly of silt with some sand and clay. They are known locally as silt bluffs, and adorn Highway 97 and the lake shore between Kelowna and Okanagan Falls. Bench levels can be correlated from one side of the valley to the other. Unfortunately, the picturesque silt bluffs of the Okanagan pose geological hazards as they occasionally fail, causing local landslides that can damage roads and buildings and be hazardous to life (see Chapter Ten).

Figure 4-9: A northerly view of Knox and Dilworth Mountains from Crawford Estates, level with the maximum height of Glacial Lake Penticton. The flatland of Kelowna was under 100 metres of water 13,000 years ago. Grouse Mountain is the peak on the left centre. Glaciation has not entirely removed the resemblance of these mountains to the towering volcanoes they once were. Tantalus Vineyards are shown in the foreground.

MOUNT DILWORTH ESCARPMENT (3)

Feldspathic lava of the Marama Formation (Figure 6-9) is exposed in a steep cliff which borders the Kelowna Golf and Country Club. The cliff was gouged out mainly by glacial action. Talus at the base of the scarp is ideal habitat for a thriving population of marmots.

FLATLAND OF KELOWNA AND RUTLAND (4)

A catastrophic flood likely struck the Okanagan whenever Glacial Lake Penticton suddenly drained. Perhaps up to 100 metres of the same material forming present day bluffs was locally removed from the valley sides. This may have occurred during one or more major water releases as the ice dam at Okanagan Falls was periodically breached and as the land rose in response to the disappearance of ice that had depressed the rocks below. The floor of that former gigantic spillway is now occupied by the peaceful and unassuming communities of Rutland and Kelowna.

Sediments deposited on the former spillway are composed of alternating layers of silt, sand and gravel with some organic-bearing clay beds. These deposits represent deltaic and shallow water deposits and marshes from runoff of Mission and Mill (Kelowna) Creeks which were far larger streams seven to eight thousand years ago. The streams flowed into a shallow bay of an early stage of Okanagan Lake. The bay gradually infilled forming a lowland on which Kelowna and Rutland were built (see Surficial Geology map, Figure 5-5). Duck Lake and Ellison Lake were likely isolated from early Lake Okanagan by similar extensive alluvial fans deposited on the valley floor by early activity of streams such as Mill Creek and Scotty Creek.

Figure 5-9: The volcanic rocks of Dilworth Mountain have been eroded by rainwater to form a series of hoodoos, visible from Knox Mountain viewpoint where this photo was taken.

Isolated knolls of sand and gravel on the northern Kelowna flatland represent the top beds of fan or deltaic deposits dissected when the last remnants of Glacial Lake Penticton drained. The gravel is utilized as aggregate and the low sand ridges provide ideal foundation conditions for small buildings.

Southward, the flatlands are composed of silt, clay and fine sand that contains plant remains and organic material such as peat. This provides evidence for a prolonged period of marsh development along the delta front in the flatlands during post-glacial time. Prior to urbanization the flatland was marked by a number of ponds representing cutoff meanders and flood channels of the various creeks that flowed across the area. Naturalists have been struggling for years to preserve the few remaining remnants of these ponds.

Figure 6-9: Feldspathic lava outcrops of the Marama Formation form this dramatic cliff bordering the fairway of the Kelowna Golf and Country Club. Glacial and ancient stream erosion combined to form this feature during the Pleistocene Period. Photo by Debra Roed.

TERRACES OF RUTLAND AND THE RUTLAND FAN (5)

Picturesque orchards blanket the terraces of the eastern part of Kelowna and Rutland. As Glacial Lake Penticton was filling, meltwater poured out of tributary valleys, the largest of which is Mission Creek valley. A huge outwash delta known as the Rutland Fan formed. As the delta grew out into the lake its sand and gravel covered clay sediments deposited earlier along the edge of the lake. These sand and gravel deposits now form the vegetable and fruit producing terraces of Rutland. As the lake drained further, Mission Creek carved even deeper terraces along its present course.

MISSION CREEK REGIONAL PARK (6)

Mission Creek Regional Park straddles present-day Mission Creek and a terrace composed partly of silt and clay deposited in Glacial Lake Penticton about 13,000 years ago, massive till (Figure 7-9) and hummocky outwash sand and gravel, capped in places with eolian sand.

Pits or kettles in the hummocky outwash terrain in the southern part of the park resulted from deposition of sand and gravel from meltwater flowing around large stagnant blocks of ice left behind as the Fraser Glacier receded. In places the kettles are occupied by spring fed ponds that teem with aquatic life (turtles are children's favourites) and provide habitat for numerous water birds. The hummocky nature of the park could also be explained by a slumping of the adjacent escarpment.

Mission Creek today supplies one third of the total runoff to Lake Okanagan, and is the main spawning channel of Kokanee. The land-locked salmon likely made their way into the Okanagan when a remnant of Glacial Lake Penticton was connected by streams to the Pacific Ocean. (Note: Perhaps ancestors of Ogopogo also migrated at this time.)

VARVED CLAY, GLACIAL LAKE PENTICTON (7)

Varved clay from Glacial Lake Penticton occurs along the north side of KLO Road just past Mission Creek bridge (Figures 8-9 and 9-9).

Figure 7-9: A hoodoo of till in Mission Creek Park overlooks the flatland of Kelowna. Dilworth Mountain is in the background. Acrylic painting by M. Roed.

These rhythmic deposits are common in glacial lakes and better examples cannot be found anywhere. The thick light-colored layers were deposited in the summer when glacial streams carried an abundant supply of silt into the lake. Thin dark layers formed during the winter when stream freeze-up curtailed the silt supply. Thus, a year's deposition is represented by one light and one dark layer. It is estimated that there could be at least two hundred varves, adding all the outcrops, totalling two hundred years of deposition as a minimum for Glacial Lake Penticton. The outcrop on KLO Road also contains 2 to 4 cm long dropstones, rocks rafted out into the lake by icebergs and then dropped when the ice melted.

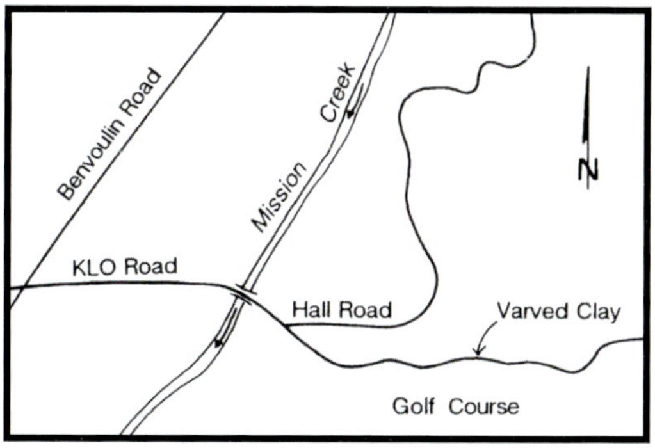

Figure 8-9: Location of varved clay along KLO Road above Hall Road intersection. There are several well exposed sections of the varved clay.

HUMMOCKY MORAINE, EAST KELOWNA (8)

The area between June Springs Road and KLO Creek valley consists of hummocky moraine, an area of hills and depressions composed of sand and gravel, and till. These deposits formed about 14,000 years ago when a lobe of glacial ice occupying Mission Creek Valley stagnated. This caused till and glacial outwash sediments to accumulate at the end of the glacier. This material eventually blocked drainage from the Mission Creek valley forming a long narrow glacial lake up Mission Creek valley and Belgo Creek valley. (Figure 5-5)

Depressions in the landscape along this part of McCulloch Road are called kettles and the hills are termed knobs, thus "knob and kettle" topography is characteristic of hummocky moraine. McCulloch Road is built on the edge of a meltwater channel from a high stage of KLO Creek that partly dissected the hummocky moraine. (Figure 10-9)

GALLAGHER'S CANYON & SCENIC CANYON REGIONAL PARK (9)

Ancient high levels of Mission Creek have carved a deep and highly picturesque canyon, Gallagher's Canyon, that extends for seven kilometres upstream. Cataracts, falls, rapids and colorful rocks of the Monashee

Figure 9-9: Varved clay and silt deposited in Glacial Lake Penticton outcrops on KLO Road. (See Figure 8-9).

Gneiss, dramatic cliffs of Eocene volcanic and sedimentary rocks, and glacial deposits combine to form this picturesque and rugged canyon. The main Mission Creek waterfall (Plate 11-9) is over 20 metres high and presents an impressive example of rock sculpturing that few people have seen.

Other geological focal points in the Gallagher's Canyon area include a 200 metre high wall of layered Tertiary volcanic rocks making up Layer Cake Hill, Pinnacle Rock (Figure 12-9), and excellent examples of folds and faults in sandstone bearing witness to the tremendous mountain building forces that have affected the Okanagan Valley (Figure 13-9)

The lower part of the canyon was the centre of an active gold mining camp in the late 1800's (see Chapter Twelve for more on gold and Dan Gallagher). Scenic Canyon Regional Park covering 75 hectares on the eastern fringe of Kelowna contains portions of this landscape that has resulted from a surprising variety of earth processes (Figure 10-9).

Layer Cake Hill

One of the most striking geologic features in the park is the layering in Layer Cake Hill, an "andesitic" lava flow or flows dated at about fifty million years old (Eocene Epoch). At least thirty two "layers" are exposed in a precipitous rock wall over two hundred and ten metres high along the northern edge of Gallagher's Canyon (Figure 12-9). The cause of the layering remains a subject of heated debate amongst geologists (see Stop 1, the Hardrock Trail, Appendix B).

Folded Sandstone Strata

On the north and south sides of Mission Creek in the western part of Gallagher's Canyon are folded beds of siltstone, sandstone and conglomeratic sandstone belonging to the White Lake Formation of Eocene age (Figure 13-9 and Figure 5-4).

These contain abundant fossil plant remains. Locally the sandstone has been moulded into symmetrical hoodoo-like sculptures, and caverns have been scooped out of the sandstone by stream erosion. Some caverns eroded into altered volcanics on the south side of the creek were used by the early Chinese gold miners as large ovens; the rocks at creek level still show the black sooty path of the smoke from these ovens.

Fault Zone

Mission Creek at one point flows along an easterly trending fault or rupture zone in the bedrock that separates fifty million year old rocks on the north side of the creek from metamorphic rocks that may be up to two billion years old on the south. This fault apparently offsets the Okanagan Valley Fault (see Chapter Four and Appendices A and B).

Glacial Deposits and Ancient Gravels

At the Canyon exit the creek has also eroded through eighty metres of glacial material revealing ancient river gravels deposited on bedrock during preglacial time or during an interglacial period (Figure 15-9).

This ancient river valley represents an entirely different drainage pattern compared to the present, and may be the only outcrop of the locally important Rutland aquifer (see Chapter Twelve). Also, the gravels may be the main source of placer gold found in Mission Creek because gold was never found in economic quantities upstream from the canyon. Gold can still be panned in the creek gravel and in bedrock fractures downstream from this ancient river bed.

Thick sequences of glacial deposits are well exposed in other spots in the Canyon and reveal details of repeated glacial advance and retreat over the last 50,000 years or so. Terraces adjacent to the stream (Figure 14-9) represent former levels of deposition along the edges of glacial lakes some 13,000 years ago. Since that time Mission Creek has rapidly carved a deep canyon through the soft glacial sediments.

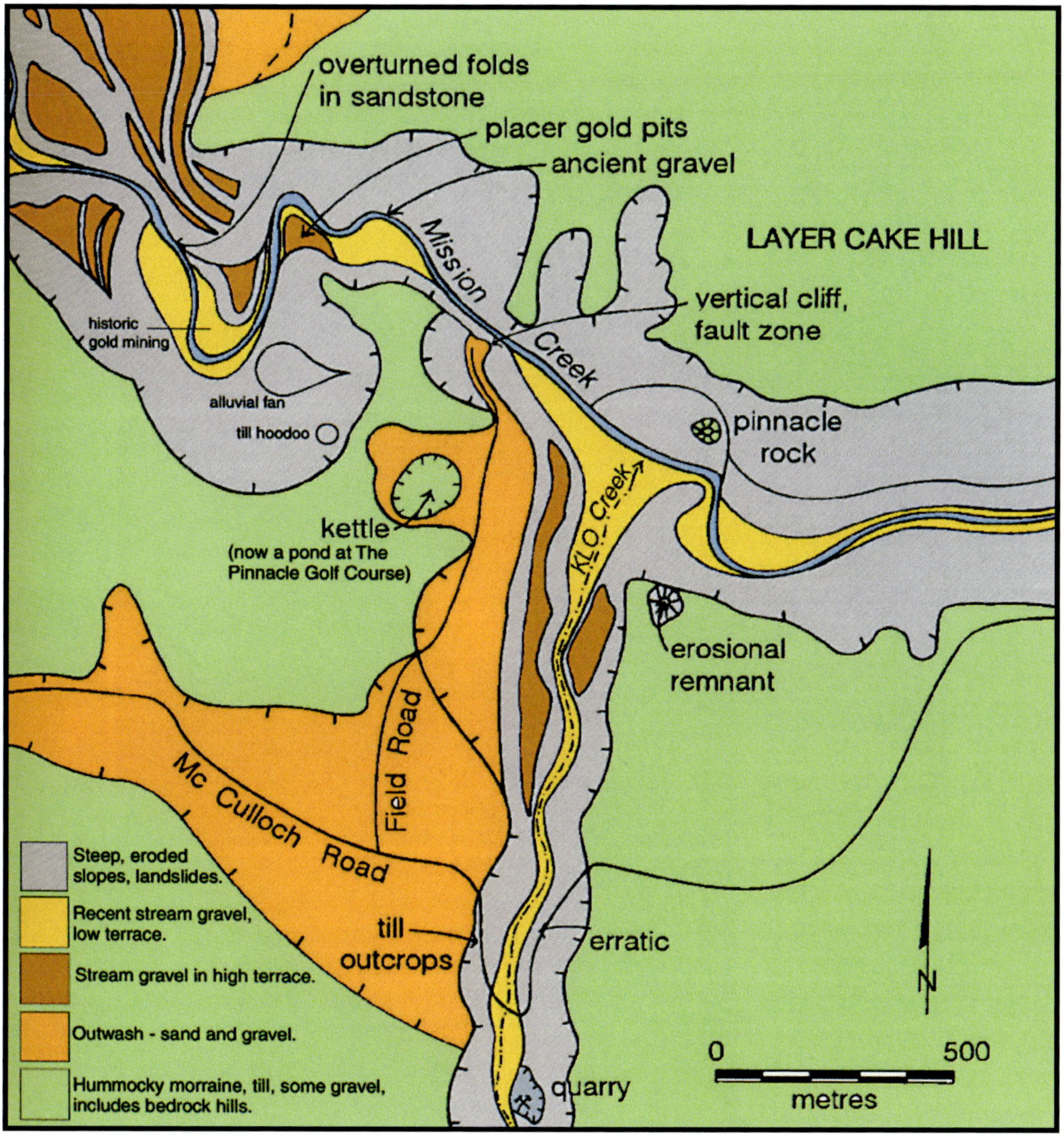

Figure 10-9: Terrain features in the vicinity of Gallagher's Canyon and Scenic Canyon Regional Park.

Figure 11-9: Few people have seen or even know about the Mission Creek waterfall (front cover) despite the fact that it occurs a short hike south of the Black Knight rodeo grounds. Youthful adventurers in the lazy sunny days of summer ride the lower part of the falls in tubes as the water plunges into a "boiling" pool. It is not easy to get a good look at the falls. It is a dangerous hike along the valley bottom with steep cliffs adjacent to deep pools of swiftly flowing water and unpredictable snags. The falls in the last 10,000 years have carved a sinuous steep cavern through hard gneisses of the Precambrian/Devonian aged Monashee Gneiss with a drop of at least 20 metres defining the entrance into Gallagher's Canyon. The water thunders and roars during the spring freshet. This is an acrylic painting of the falls by M. Roed.

Figure 12-9: A westerly view of the curious layering of volcanic rocks making up Layer Cake Hill. A possible explanation for this is given in the field trip "Hardrock Trail" in Appendix B.

Pinnacle Rock

Pinnacle Rock, a curious spire-like landform near the base of Layer Cake Hill (Figure 9-12) is composed of the same layered volcanic rock as the Hill itself.

Although Pinnacle Rock may be an erosional remnant its formation is open to question. Aerial photos show that locally there are gaping vertical fractures in the hill. It is therefore possible that Pinnacle Rock slipped off the side of the escarpment as a block glide feature during downcutting of the valley by Mission Creek. Alternatively, Pinnacle Rock may simply be an erosional remnant carved out by deglaciation meltwater plunging into Mission Creek from KLO Creek.

TILL OUTCROPS; THE DESCENT INTO KLO CREEK (10)

Just past Field Road, McCulloch Road descends into the KLO Creek Valley (Figure 17-9). Outcrops of massive gray glacial till in the road cut are amongst the best in the area (see also Figure 6-9). These poorly sorted and unstratified sediments contrast sharply with most glacial deposits in the Kelowna area which tend to be water-lain silts or water-worked sands and gravels.

QUARRY, MONASHEE GNEISS (11)

The oldest group of rocks in British Columbia are well exposed in a quarry (see Figures 10-9 and 18-9)) at the bottom of KLO Creek valley. The distinctly banded rock in the quarry walls is mylonite, a high grade metamorphic rock making up part of the Monashee Gneiss (Figure 16-9).

Different coloured bands in the rock contain different minerals. Dark bands are enriched with a mineral known as hornblende, and lighter bands contain an abundance of feldspars such as microcline. Notice the stretched quartz veins in many samples on the quarry floor. Structures such as these indicate that the rocks have been subject to tremendous forces.

Figure 13-9: Folded and overturned sandstone of the White Lake Formation in Gallagher's Canyon provides insight into the mountain building forces that once affected the Kelowna area.

The layering and segregation of these minerals were caused by extreme heat and pressure which completely changed the original rock type. It is these features of the rock that indicate to geologists that the rock has been deeply buried. In fact, the current interpretation is that these rocks were originally granite and formed the edge of the early continental core of North America. Movement along the Okanagan and Mission Creek Faults since Eocene time brought them to the surface.

EAST KELOWNA FAN (12)

The peaceful orchards, hobby ranches and equestrian centres of East Kelowna are situated on terrain similar in origin, though formed slightly earlier than the Rutland Terraces. The first phase of meltwater coming out of Mission Creek valley, after diversion around Layer Cake Hill, formed a delta in East Kelowna (see cross section, Figure 1-4, Chapter Fourteen). As Glacial Lake Penticton drained and the once ice-laden land rose in response to unloading, Mission Creek dissected the delta. Eventually the creek established its present channel to the north leaving a series of near-dry valleys such as "Gully Road", marking the courses of these early stream channels in East Kelowna.

BEACH DEPOSIT, CRAWFORD ESTATES (14)

Crawford Estates is situated on a high outwash gravel terrace marking the maximum height to which Glacial Lake Penticton rose. The only well developed Glacial Lake Penticton beach deposit in the Kelowna area occurs on the rim of this terrace at a site now designated as a Natural Area by the City. The deposit can be viewed to the west of Westridge Road (beach-front property) overlooking a large rehabilitated gravel pit which exposes the internal structure of the wave-built beach deposit (Figure 19-9).

Part of this sand deposit was reworked into dunes by eolian(wind) activity. This is an important geologic landmark and could hold evidence of the first human activity in the Okanagan Valley (e.g. charcoal from camp fires or stone implements and tools may form middens buried beneath the sand). Unfortunately part of the locality is still destined for development.

Other features in this area include Crawford Falls (30) and Bellevue Creek Canyon (29).

Figure 14-9: An acrylic painting by M. Roed showing the steep rock walls of Gallaghers Canyon and volcanic layers of Layer Cake Hill, the same rock wall shown in Figure 3-3.

Figure 15-9: Rusty gravel outcrops at the exit of Gallagher's Canyon. It may be the only outcrop of the Rutland aquifer in the area, and the source of most of the gold in the creek flats below.

Figure 16-9: Gneiss belonging to the Shuswap Metamorphic Complex or Monashee Gneiss, the oldest rock in British Columbia. Also, see Stop 2 of the "Hardock Trail" field trip, Appendix B. photo by Debra Roed.

LAMBLY CREEK BASALT, LAKEVIEW (15)

For about two kilometres south of the floating bridge, on the west side of Okanagan Lake, are lava flows belonging to the 762,000 year old Lambly Creek valley basalt (Figures 8-4 and 9-4). At least three separate flows outcrop in the steep escarpment paralleling the lake. Locally the basalt contains white calcareous fragments, apparently representing soft sediment picked up by the lava as it flowed. The sediment could hold valuable information on Okanagan climate at the time of basalt extrusion.

KALAMOIR REGIONAL PARK (16)

The narrow road down to Kalamoir Regional Park, accessed from Collens Hill Road, affords close up views of dacite, a kind of volcanic rock that also makes up much of Mt. Boucherie. At this location the dacite forms columns or structures similar to those commonly observed in basaltic lava flows. Outcrops of conglomerate near the bottom of the road are representative of the White Lake Formation.

MOUNT BOUCHERIE COLUMNS (17)

Mount Boucherie is likely part of a former lava dome that dominated the landscape 50 million years ago. It has been carved and moulded by repeated glacial advances but remains a prominent geologic landmark in the Kelowna area. High on its eastern flank the mountain is seen to be made up of folded volcanic flows composed of dacite (Figure 22-9) The rocks also display columnar structure.

Figure 17-9: Little White Mountain, the highest point in the Kelowna area, provides a backdrop for the Kettle Valley Railway that winds along the edge of the upland. The view is from the Scotty Creek area looking south-southeast.

The forest below the mountain is littered with large columnar blocks of dacite, resembling scattered ruins of an ancient temple and a forgotten civilization (see Figure 1-A, Appendix A). Recent mapping by the authors has revealed new details about Mt. Boucherie given in Appendix C.

SANDSTONE CAVITIES (18)

A spectacular cliff of White Lake Formation conglomerate occurs at the junction of Boucherie Road and Old Boucherie Road (Figure 7-B). Some parts of the 20 metre high cliff have been preferentially eroded producing numerous cavities. Fresh rocks at the base of the cliff show that the differential weathering is the result of oval areas in the conglomerate, known as concretions, being cemented together with the mineral calcite. Quartz cement holds the remainder of the rock together. Over the centuries mildly acidic rain water running over the outcrop has dissolved away the calcite allowing the contained boulders and pebbles to fall out of the cliff (see also Stop 5, the Hardrock Trail, Appendix B).

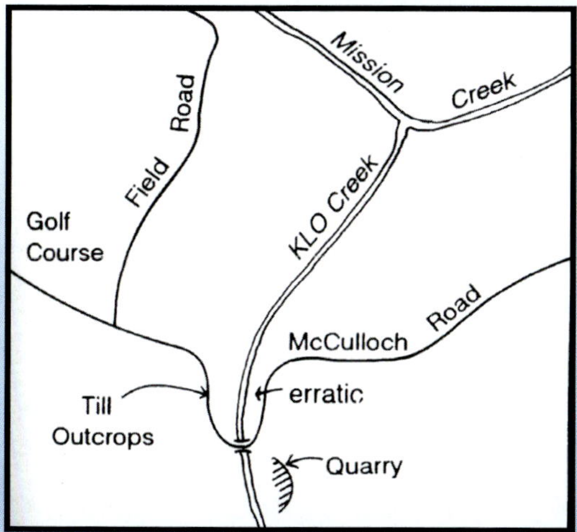

Figure 18-9. Location of Quarry, Till Outcrops and Erratic along McCulloch Road.

Figure 19-9: Looking like a desert, sand representing a beach at the highest level of Glacial Lake Penticton parallels Westridge Drive in Crawford Estates. Hayman/Macdougall Ridge and Okanagan Lake are in the background.

Oil Well Fiasco

In the early thirties the Okanagan Oil and Gas Company Limited announced plans to drill for oil at Kelowna. The promoters raised money from local businessmen to drill a series of wells just off Lakeshore Drive where it crosses Mission Creek. At least one hole was drilled through unconsolidated glacial sediments to a depth of nearly 1,000 metres, finally hitting a granitic type of rock (probably the Monashee Gneiss). Investors are still waiting for the oil to be turned on.

Figure 20-9: Early photo of the derrick in 1931 showing the Kelowna oil well with all of the shareholders anxiously waiting for the oil to be turned on

Number 872 · Shares 60

Incorporated under the laws of the Province of British Columbia

Okanagan Oil and Gas Company Limited

Registered Office · Kelowna B.C.

200,000 shares of $1.00 each

This Certifies That Ena Rose Stewart of R.R.1 Summerland BC is the registered holder of Sixty Shares fully paid up and non-assessable numbered 162637 to 162696 inclusive in the Capital Stock of

OKANAGAN OIL AND GAS COMPANY LIMITED

subject to the Memorandum and Articles of Association of the Company transferable only on the books of the Company by endorsement hereon and surrender of this Certificate.

In Witness Whereof the said Company has caused this Certificate to be signed by its duly authorized officers and to be sealed with its Corporate seal at Kelowna BC this 12th day of August A.D. 1932

Director · Secretary

$1.00

Figure 21-9: A share certificate of the Kelowna Oil and Gas Co. Ltd., proving there were many optimists in Kelowna. A news article dated Thursday, September 3rd, 1931, from The Kelowna Daily Courier and Okanagan Orhardist is also reproduced here, along with a photo of the “Kelowna Oil Well”. These items were supplied through the courtesy of the Kelowna Museum.

THE KELOWNA COURIER

AND OKANAGAN ORCHARDIST

Kelowna, British Columbia, Thursday, September 3rd, 1931

BORE OF OIL WELL BELOW SEA LEVEL

Indications Are Promising And Natural Gas Is Expected Within Next Eight Hundred Feet

Drilling was resumed at Kelowna Oil Well No. 1 on August 25th, the day after the casing was set, and a depth of 1,770 feet has now been attained. Sea level was passed at 1,737 feet. Mr. Julius Rickert, the geologist of the Okanagan Oil & Gas Co., Ltd., has waited patiently for sea level to be reached, and the formations encountered at that point verify his prediction and strengthen his opinion that natural gas will be encountered at about 2,500 feet or less.

There are considerable showings of gas, as well as oil seepage, and it is confidently expected that, on account of the greater depth, fewer troubles will be met in drilling, owing to lessening of possibilities of water and caving conditions, the rock formations now being more stratified.

The dogged perseverance shown by the founders of the enterprise in the face of many disheartening setbacks and delays is arousing keen interest and sympathy throughout the province and the Okanagan in general, as it is being recognized that genuine effort is being made to prove for once and for all the oil and gas possibilities of a formation which is not confined to the Kelowna district and therefore is of vast importance to the province in general. It is to be hoped that the sympathy will be translated into material terms of cash support, without which the undertaking cannot be carried to a conclusion. So far, the people of the Kelowna district and of other communities in the Okanagan have given a fair measure of assistance, but further funds will be required to make a thorough test, and any appeal made by the Company should meet with a liberal response.

MCDOUGALL CREEK FAN (18)

McDougall Creek alluvial fan dominates two kilometres of the west shoreline of Okanagan Lake below Mount Boucherie (Figures 5-5 and 1-9). The fan was deposited in the early Holocene perhaps seven to eight thousand years ago in response to extensive erosion in the McDougall Creek watershed. This potentially powerful stream has in the past moved large boulders and formed hazardous "debris flows" that today would cause extensive damage. There remains some potential for flooding during extreme climatic events, although recent remedial measures in 2013 have reduced some risks.

VARVED CLAY, BOUCHERIE AND GELLATLY ROADS (19)

The deep roadcuts along Boucherie Road and on Gellatly Road between the lake and Highway 97 at Westbank allow more views of varved clay formed in Glacial Lake Penticton. This terrace correlates with the same lake deposits as found at Kelowna.

GLEN CANYON REGIONAL PARK, POWERS CREEK, GLENROSA AREA (20)

Although it is a steep trail, the climb down into Glen Canyon Regional Park is rewarding (Figure 23-9). Glacial outwash escaping down Powers Creek valley eroded a steep walled 100

Figure 22-9: Columnar structure is well developed at the northern summit of Mountain Boucherie. The rock is a dacite, similar to basalt and belongs to a massive vertical pipe that cut through an older part of this volcano. Another volcano, Black Knight Mountain, is visible in the upper right of this acrylic painting by M. Roed. The burnt trees shown are the result of a forest fire in the early evening of May 7, 1992. It was accidentally started by two 11-year-old boys playing with matches. The fire consumed 60 hectares of forest.

metre canyon through massive conglomeratic beds of the White Lake Formation (Figure 7-4). Locally, thin coal beds can be seen in these ancient deltaic-like deposits along the canyon walls. Since deposition the rock layers have been tilted by tectonic forces and thus are no longer horizontal.

COAL BEDS, WHITE LAKE FORMATION (21).

Coal outcrops as thin beds in the White Lake Formation across from Gorman Bros. Lumber Ltd. lumber mill. (See Stop 4, Hardrock Trail field trip, Appendix B).

KELOWNA BOG (22)

The Kelowna bog is located just south of the airport. This is where Alley (1976) first described the post-glacial history and climatic and vegetation change in the Okanagan Valley.

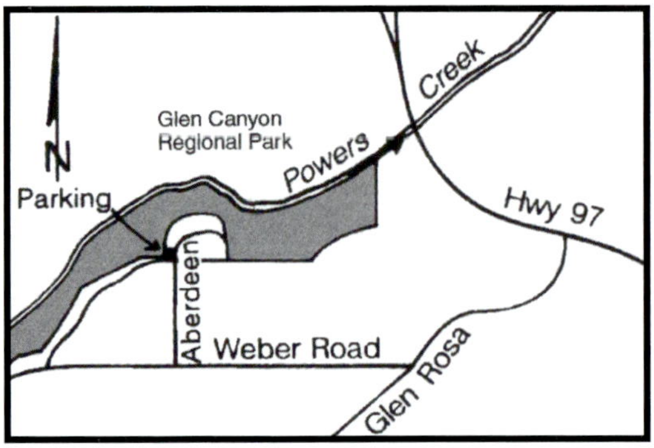

Figure 23-9: Location Map of part of Glen Canyon Regional Park.

FLUTED MORAINE AND DRUMLINS (23)

Linear ridges of moraine are easily visible along Okanagan Valley east of the airport in the vicinity of the Sunset Ranch golf course (Figure 24-9). These features formed at the edge of the Fraser Glacier as it flowed along the valley. Many are over one kilometre in length and are locally cut by dry valleys in places representing former meltwater channels. The ridges are called fluted moraine and form from moulding at the base of a flowing glacier. Drumlin-like landforms, carved into rock at the base of a glacier, are common in the bedrock of Okanagan Mountain Park. Examples can be seen looking across the lake from Highway 97 at Peachland (Figure 26-9).

BLACK KNIGHT MOUNTAIN (24)

Black Knight Mountain is a giant roche moutonnée, carved and plucked at the base of a southeastward moving glacier. The mountain is made of volcanic rock related to the same volcano that forms Layer Cake Hill. It is one of the most prominent landforms in the valley (Figure 25-9).

ESKERS AND KAMES (25)

An intricate group of ice-contact landforms known as eskers and kames occur in the Black Mountain area south of Highway 33, in the vicinity of the rodeo grounds. Most of these deposits have been mined for gravel.

Figure 24-9: Long linear ridges of moraine form distinctive features of the Okanagan Valley near Sunset Ranch golf course. These ridges are fluted moraine, moulded at the base of the Fraser Glacier.

Figure 26-9: Hard gneissic bedrock of Okanagan Mountain Park displays streamlined glacially abraded landforms (rock drumlins) in this eastward view near Peachland.

Figure 25-9: Black Knight Mountain has been molded by an overiding glacier into a giant roche moutonnée. This view shows the western side of this local landmark.

History of the Kettle Valley Railway

A thin line through the forest far up on the hillside overlooking Kelowna is the only visible suggestion of the right-of-way of the Kettle Valley Railway. It was built by the Canadian Pacific Railway Company, and others, during the period 1901 to 1916. Construction was under the supervision of Andrew McCulloch (born in 1864), a remarkable Canadian engineer. The line operated from 1916 to 1962, hauling freight and resources between the Kootenays and the Pacific coast. It records a most intriguing and interesting saga of the development of British Columbia, a "magnificent irrationality" to quote Beth Hill, a local historian. Now abandoned, it is the focus of a strong and determined group of railway buffs, historians, hiking enthusiasts and others who are aiming to preserve its 500 kilometre corridor as a biking, skiing and hiking linear park.

The right-of-way of the Kettle Valley Railway, nearly 1,000 metres above Kelowna, can be accessed by taking the Myra Forest Access Road off McCulloch Road, or by following June Springs Road in East Kelowna. Between these two points the Kettle Valley Railway offers spectacular views of Kelowna and Okanagan Lake and impressive panoramas of deep canyons such as the two forks of KLO Creek, formerly known as Canyon Creek, and historically referred to as Myra Canyon. This section along the old grade contains sixteen high trestles that are engineering masterpieces, and two rock bound tunnels one of which is over 100 metres long. The largest trestle at Bellevue Creek is of steel construction 220 metres long and 55 metres high, but most are wooden. Walking or biking across these chasms can be an unnerving and chilling experience. Rock tunnels are somewhat hazardous. Rehabilitation of the trestles by community service groups and a great number of benefactors started in 1993. If one is cautious, this section of the KVR is perhaps one of the most enjoyable outings in the whole Okanagan Valley. The Okanagan Mountain Park Fire of 2003 destroyed 14 of 18 trestles. Despite this, details of the geology along the main section of this railbed are given in Appendix C in expectation of eventual restoration, which has happened.

Figure 27-9: First KVR passenger train at Penticton, May, 1915.

CHAPTER 10

GEOLOGIC HAZARDS IN THE OKANAGAN VALLEY

by Murray A. Roed

Geologic hazards are natural processes that can be damaging to private property, structures and/or human life. Five major types include landslides, sinkholes, floods, earthquakes and volcanic eruptions.

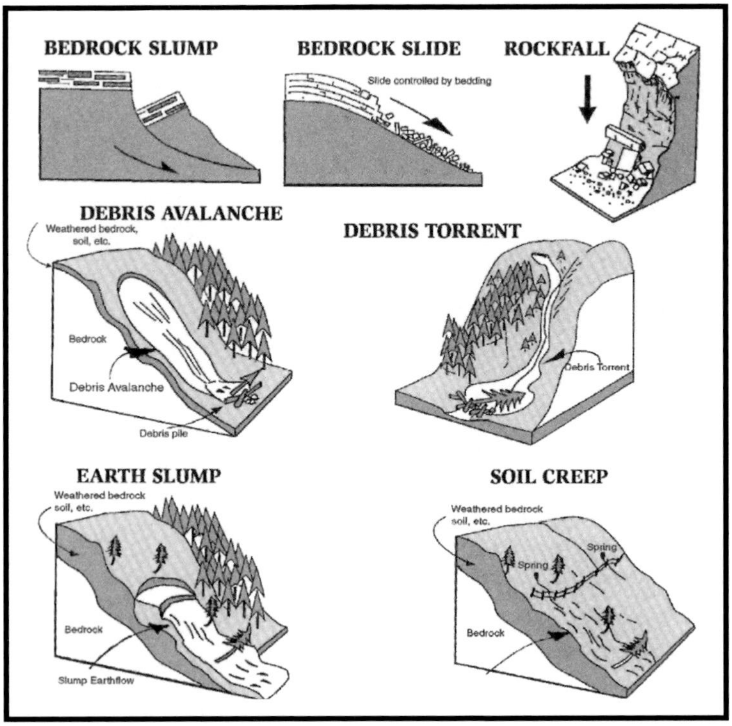

Figure 1-10: Some Landslide Types (after Chatwin et al, 1994).

LANDSLIDES

Landslides are facts of life in British Columbia, and the Okanagan is no exception. Landslide is a general term used to describe the down-slope movement of soil, rock and organic materials, wet or dry, and the landform that results. There are various types of landslides; main types are shown in Figure 1-10.

Figure 2-10: This is a stereo-pair of the Mission Creek rotational slump nine kilometres east of Kelowna and just south of Highway 33. A stereoscope is required to see the terrain in 3-dimensions. (Air photos 30BCC 92092 – 3 & 4 reprinted courtesy of Province of British Columbia, Surveys and Resource Mapping Branch.)

CAUSES OF LANDSLIDES

Landslides are often caused by more than one factor. Undercutting of a slope by a river is the most common. This was the main cause of a spectacular rotational slump which occurred in October 1984 on Mission Creek nine kilometres east of Kelowna along Highway 33 (Figure 2-10).

Approximately 500 metres of the north bank of the creek valley failed and completely blocked Mission Creek. The thunder clap sound produced during failure was heard as far away as Orchard Park shopping mall 10 kilometres to the west (Don Dobson, personal communication). This landslide involved over ten million tonnes of material and nearly destroyed a portion of Highway 33. It is an important geologic landmark and a vivid reminder of the hazard that exists on some Okanagan slopes.

Another cause of landslides is the addition of water either from prolonged rainfall, melting snow, or groundwater seepage. Water pressure builds up in sediment pores causing a loss of strength and, ultimately, failure. Terrain underlain by silt and clay of Glacial Lake Penticton is particularly sensitive to pore pressure build-up along slopes. Many slopes in the Kelowna area, such as near the base of Knox Mountain, along the Rutland Terrace and along Swamp Road are characterized by earth flows and slumps. For example slumping occurred on a terrace at the northern end of Gordon Drive on August 21, 1993 (Mountain Avenue slide). Three houses were damaged when water-saturated silt and clay extruded at the base of a terrace slope; backyards of two houses dropped six to twelve metres as a result (Figure 3-10).

Landslides have been caused by the unintentional or accidental release of excess irrigation water on orchards above steep slopes. This was the apparent cause of a slide on Lakeshore Drive at Summerland during 1992 and in another location here in 1970 which destroyed three homes and killed one person (Evans, 1982). Over a dozen large slides have occurred along Okanagan Lake since 1930 and some along Osoyoos and Skaha Lakes. Large silt slumps still occur unpredictably along Highway 97 especially between Summerland and Penticton.

Earthquakes are the most widely known cause of landslides. Examples can be cited for many parts of the world. Although Kelowna has a low earthquake risk, whenever a slide occurs investigators always check to see if it may coincide with the time of a recorded earthquake.

Clearcut logging and forest access roads contribute to the probability of landslide activity in some localities, although forest companies are working hard to reduce this hazard. In upland clearcut areas more rainfall seeps into the ground which in turn may increase pore pressure on adjacent slopes as water tries to escape. These conditions may persist until the clearcut has hydrologically regenerated. Surface water runoff also increases in clearcut areas potentially raising the extent of erosion, undercutting of banks, and the probability of debris transport during rain storms, especially if access roads are not carefully constructed, maintained or deactivated.

Figure 3-10: This person's backyard and patio on Mountain Avenue at the end of Gordon Drive in Kelowna took a "down elevator" the night of August 21, 1993. No one was injured.

THE ROLE OF GROUNDWATER

The influence of groundwater on slope stability is not commonly appreciated by the general public. The addition of water to slope-forming materials has two major consequences. First, the soil becomes more buoyant; this reduces the strength of the soil. Second, depending on the pressure imposed by the groundwater build-up in combination with the type of soil on the slope and the actual slope angle, buoyant forces and pore pressure may combine to overwhelm the natural resistance of the soil to gravity and a failure occurs. In some cases the failure takes the form of an explosion, like the popping of a champagne cork. Even if the pop-out on a slope is small, the resulting small debris flow can trigger a much larger failure downslope, or it may "snowball" into a larger one. The former happened in the Belgo Creek debris avalanche in the Joe Rich District slide in June of 1990. Runoff from a poorly maintained forest access road drained into a sensitive part of a slope that was naturally somewhat unstable.

THE ANATOMY OF A DEBRIS AVALANCHE

The Belgo Creek debris avalanche thirty kilometres east of Kelowna occurred on June 12, 1990, in a poorly to moderately drained till mantled depressional swale on the western slope of Belgo Creek valley. The landslide resulted in a water-charged debris avalanche over 1.5 kilometres long. Seventeen terrain units defining the slide have been recognized (Ministry of Forests, 1990, and Cass et al, 1992). Five debris torrents also occurred in the general area (Figure 4-10).

The main avalanche developed from three small related debris flows on a steep water saturated slope. These minor events and accompanying tree falls in turn triggered the main flow, the head scarp of which had received an unusual amount of runoff due to disrupted drainage on a forestry access road.

A forty-two day period preceding the mass movement events was the wettest on record. An intense storm within 24 hours of the slide combined with snow melt is believed to have created saturated conditions in the surficial materials and high pore pressure in both the surficial material and underlying fractured bedrock. This caused an explosive outburst of surficial materials and loose bedrock where slopes are in the order of 65-75% (30 to 40 degrees) and springs emanated from the bedrock. Spray from the top flow scar which is only 1.5 metres in diameter, splattered nearby trees three metres up their trunks.

Velocity of the main slide has been estimated at 10 metres per second and involved 23,000 cubic metres of soil, forest debris and a large quantity of water (Cass et al, 1992). A house was demolished and three people lost their lives (Figure 5-10).

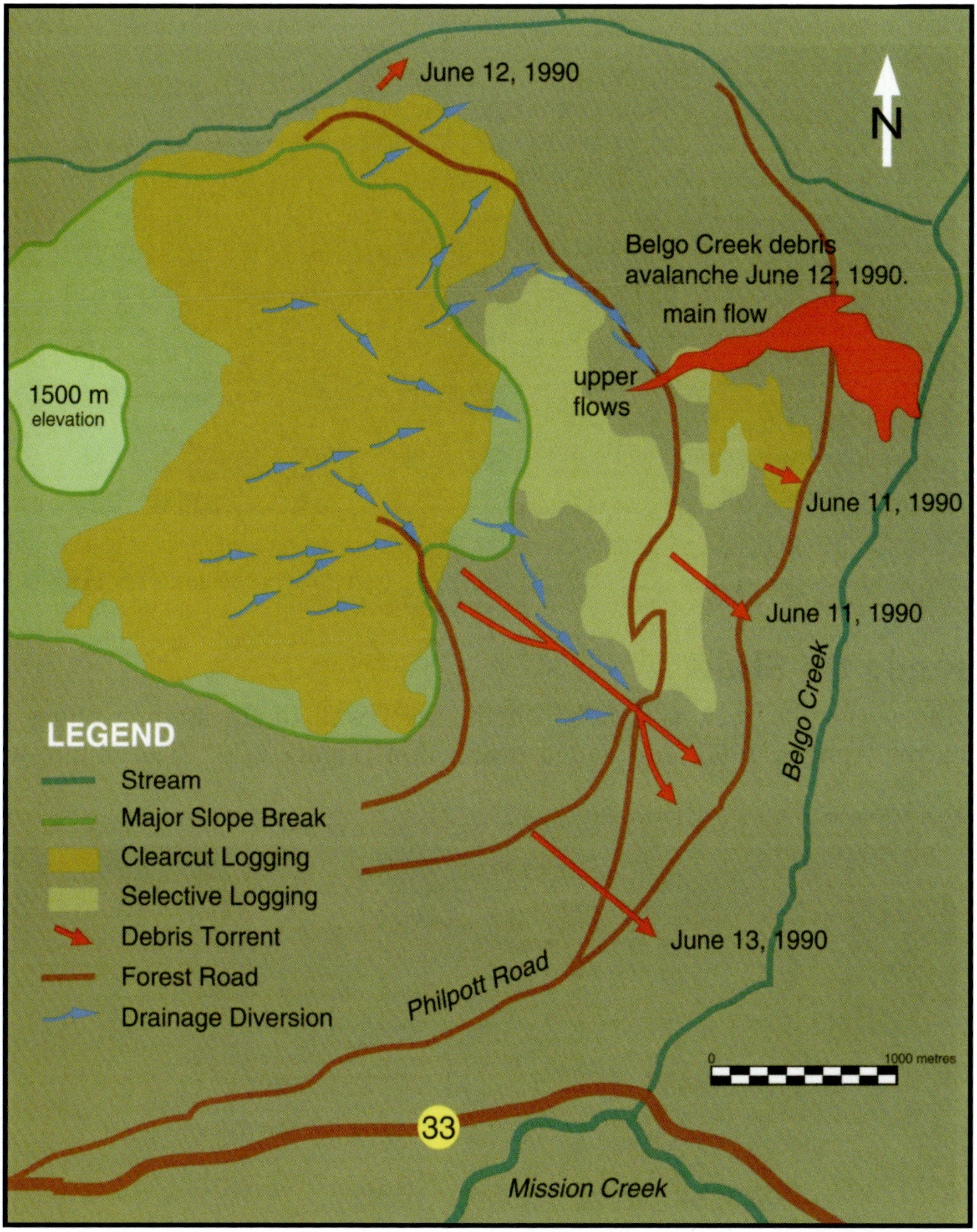

Figure 4-10: Map of the Belgo Creek Debris Failure Area, Philpott Road, Joe Rich District.

BRENDA PIT SLIDE

A spectacular rock avalanche on the west wall in the Brenda Mines pit occurred April 9, 1990, and ended production (Figure 6-10). The failure was expected and had been closely monitored for months. Approximately eight million tonnes of rock filled the bottom of the pit to a depth of over 30 metres. Cause of the slide was structural failure and high groundwater pressure on the 300 metre high wall with an average slope of about 80% (40 degrees). The slide caused a small earthquake, recorded at the Pacific Geoscience Center in Sydney, BC.

Figure 5-10: Part of the debris at the base of the Belgo Creek debris avalanche is shown here. The debris pile includes a demolished house in which three people lost their lives.

Figure 6-10: A rockslide on the west wall of the pit at Brenda Mines Ltd., shown here, terminated production in 1990. Photo courtesy of Brenda Mines Ltd.

LANDSLIDE RISK

The public is generally unaware of landslide risk areas which is unfortunate. Statistically, large and infrequent landslides contribute less to personal and property damage than smaller more frequent landslides in populated regions. Increased public awareness and more frequent consultation with qualified geotechnical personnel prior to development could limit activities that result in the undercutting of slopes, eliminate the placement of fill on steep slopes, emphasize the necessity of proper drainage, and prevent building construction near unstable slope crests.

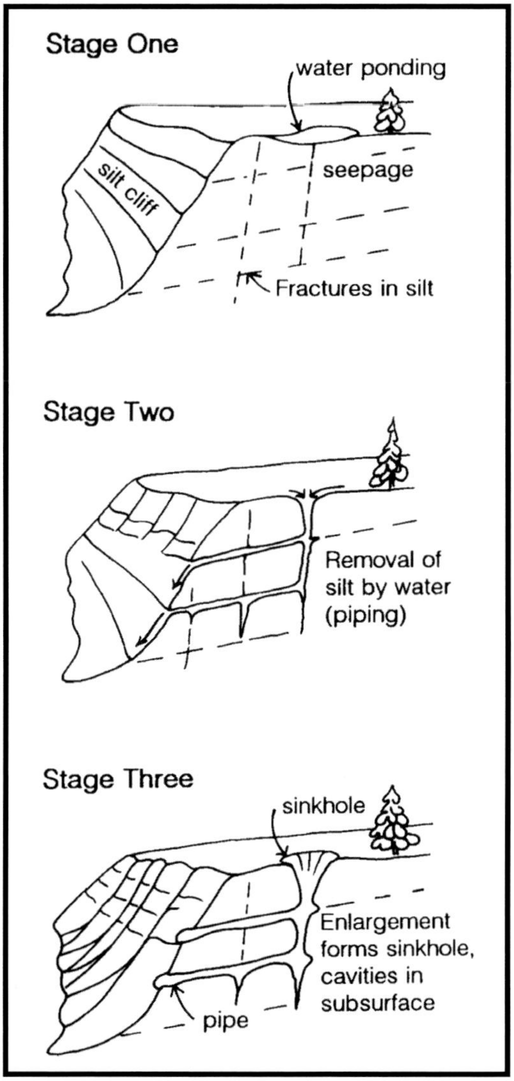

Figure 7-10: Sketches showing stages in the formation of sinkholes by piping in silts of glaciolacustrine sediment.

SINKHOLES

Sinkholes in the Okanagan result from a process known as piping. Piping occurs when water seeping along vertical and/or horizontal fractures in silt interacts with the tiny sediment particles causing a collapse of the structure holding the silt particles together. The silt particles become mobilized by water along a thin tubular pipe where the fracture originally occurred. With time the pipe enlarges, caving occurs and the ground collapses due to the extrusion of silt along the pipe. Sinkholes are common in the bluffs of Glacial Lake Penticton sediments due to their unconsolidated silty composition (Figure 6-10). They are especially well developed in the Sage Mesa area near Penticton.

Sinkhole formation destroys foundation conditions. Any ground depression underlain by Glacial Lake Penticton strata should be thoroughly analyzed before construction. Filling sinkholes is not a long term remedy. The hazard can be considerably increased by uncontrolled irrigation or even backyard watering.

FLOODS

Flooding of property in Kelowna, although commonly attributed to "too much rain" often has more complicated origins of both human and natural contrivance. Much of the city is built on a flat delta that emerged above lake level over the last part of the Holocene. In many places the water table is less than a metre below the surface and can be higher when runoff into Okanagan Lake exceeds the controlled outlet capacity at Penticton and Okanagan Falls causing the lake and water table levels to rise. In addition the Kelowna flatland is a huge groundwater discharge zone, receiving water that infiltrates from the surrounding highland. During rain storms water level is easily affected first by runoff and then by a rise in the groundwater table. This results in flooding of septic tile fields, basements, and in extreme cases, overtopping of stream channels.

Mission Creek supplies one third of the annual runoff to Okanagan Lake and has a notorious past history for flooding. The stream has now been so channelized (Figure 8-10) across the flats to prevent water from breaching the banks, that even the Kokanee may be finding the creek somewhat unrecognizable. Urbanization masks the natural sinuous channels the stream used to follow less than two hundred years ago and to which it would return to without human intervention. Several reservoirs have been constructed in the watershed (see Chapter Thirteen) and help control runoff. Thus the flooding threat is presently diminished but Mission Creek may strike again.

Like Mission Creek all streams entering the valley have alluvial fans or deltas associated with them (e.g. Fintry Creek, McDougall Creek, Trepanier Creek, Scotty Creek, etc.). These areas may be at risk from flooding particularly during a climatic extreme event, and unfortunately tend to be urbanization sites. The July 1993 flash flood on Scotty Creek damaged residences and the Sunset Ranch Golf Course illustrates one type of hazard. This flood was caused by failure of an abandoned beaver dam high up in the watershed!

The public can hardly be aware of all potential flooding hazards. However, watershed users have a responsibility to know of watershed sensitivities, to monitor activities and carefully plan upstream terrain development. It is important to ensure minimal risks to downstream facilities and people, and strive to maintain water quality.

Inadequately designed or maintained storm drainage systems can also cause "unexpected" floods. Urbanization creates impermeable surfaces such as paved streets that, during storms, increase the availability and speed of runoff water and locally elevates soil infiltration. Hazards develop due mainly to enhanced surface erosion. Ponding and infiltration leads to increased subsurface pore pressure and landslides. In addition, runoff from urbanized areas carries far more pollutants to lakes and streams than water handled by sewage treatment plants. Preservation of our surface water and groundwater quality requires new ways of dealing with storm water drainage, and a policy for retention basins in new subdivisions is now being implemented (Dayton & Knight, 1989).

EARTHQUAKES

Earthquakes are a potential hazard in British Columbia. They are caused when stresses build up in the earth much like what happens when an elastic band is stretched. Eventually the earth's crust suddenly ruptures forming a fault, or break in the crust, and instantaneously releasing the stored energy. The energy travels

Figure 8-10: Mission Creek on a summer day is a placid pleasant stream. However, it drains one of the largest watersheds in the Okanagan Valley and supplies one third of the annual runoff to Okanagan Lake. Its potential for flooding should not be underestimated.

through the earth as waves similar to the waves produced by a pebble dropped into water. These waves are powerful and may cause widespread destruction.

Plate tectonics helps us understand why these stresses build up in the earth's crust and why British Columbia is at risk. Figure 2-4 shows the principal tectonic "plates" on the Pacific coast. The plates are colliding, sliding past or moving away from one another and it is near these plate boundaries where most earthquakes originate. Plate movement is very slow, typically only a few millimetres to centimetres a year but over time energy can accumulate, and its sudden release may cause a mighty earthquake and possibly a tsunami.

KELOWNA EARTHQUAKE RISK

Scientists believe that coastal British Columbia may be overdue for a major earthquake. Much of the earthquake hazard for the Okanagan comes from the potential in the Lower Mainland or coastal zone (Ernie Naesgaard, personal communication). The distribution and intensity of earthquakes in the Okanagan shown in Figure 9-10 illustrate that the hazard potential for the Okanagan is much less than for the Lower Mainland. This reflects our distance from the plate boundaries in the offshore (Figure 2-4). For example, a 1946 earthquake on Vancouver Island registered 7.3 on the Richter Scale. It produced a high level of shaking capable of moderate building damage and set off landslides for 200 kilometres around its origin (epicenter) near Courtenay. Kelowna, 400 kilometres away, may have experienced a very low level of shaking during this event.

The potential for a major earthquake originating in the Okanagan appears very low. However, no specific work has been done but numerous scientists have examined sediment of Holocene age (the last 10,000 years) in exposures throughout the Okanagan Valley, and there is no report that these sediments have been offset by movement along a fault. Earthquake activity in historical times in the Okanagan has been restricted to minor shaking with earthquake magnitudes in the range of 2.0 to 4.5 Richter, as shown in Figure 9-10. Nevertheless it is sobering to realize that the Okanagan Valley Fault zone with numerous related faults is a major zone of weakness in the Earth's crust, and runs beneath the Okanagan Valley.

Although some very loose sediments in the Kelowna area could liquefy even during very small disturbances. It would take an earthquake with a magnitude of 7.5 Richter at Vancouver to cause potential liquefaction of most liquefaction-prone sediments at Kelowna. The estimated frequency of such an event affecting the Okanagan may be once every ten thousand years on the basis of existing geologic evidence. This is important because Okanagan communities cannot afford to have any shaking going on. Shaking causes water saturated silts and fine sands to liquefy. Kelowna and Vernon largely sit on these sediment types. The potential for liquefaction is one reason why engineers did not build a suspension bridge across Lake Okanagan; a floating bridge was constructed instead at a much lower cost.

The probability may be very low but the potential ramifications of high level shaking are daunting. Depending on site conditions, buildings on the flats of Kelowna especially could be damaged or destroyed, houses would settle or sink, landslides could be triggered.

A large earthquake centred in the Okanagan Valley could generate giant destructive waves (called seiches) in Okanagan Lake causing havoc and probably loss of life along its heavily populated shore. Fortunately, the possibility of this appears extremely low.

EARTHQUAKE PREDICTION

Earthquake prevention is presently impossible as is accurate prediction. However, the destructive consequences can be reduced by proper planning, appropriate building site selection, structural design through geologic and geotechnical studies, an informed public, and a prepared emergency response team.

VOLCANIC ERUPTIONS

Major periods of volcanic eruption that affected the Okanagan in the geologic past were described in earlier portions of the book. Although there are three different active volcanic belts within British Columbia that produced over 100 eruptions during the Quaternary Period, the Okanagan area has been largely, though not entirely, inactive and minimally affected.

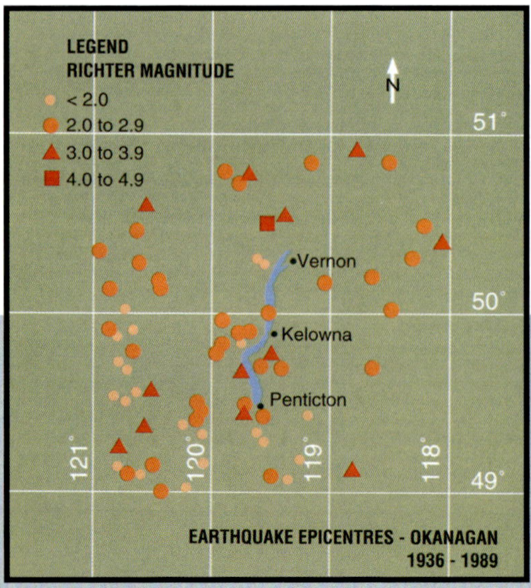

Figure 9-10: Compilation of Earthquakes in the Okanagan from 1936 to 1989. The largest was northwest of Vernon in 1936 at 4.5 Richter (Data courtesy of Pacific Geoscience Center, Sidney, British Columbia). The Richter scale is gradually being phased out of use and replaced with a new scale based on magnitude ("M" scale) of energy released. The two scales are very similar for quakes less than 7.0.

VOLCANIC ASH FALLS AT KELOWNA SINCE THE ICE AGE

Local surficial sediment deposits formed over the last 10,000 years contain at least four layers of volcanic ash. The oldest is the Mazama Ash derived from an explosion that formed Crater Lake in Oregon 7,600 years ago (Figure 10-10). Just south of the Kelowna airport the Mazama Ash is over one metre thick where it accumulated in a depression [(22) in Figure 1-9, see also Figure 1-6]. Imagine the environmental impact of even a fraction of that ash fall would have on Kelowna today, and this from a source 700 kilometres away! Ash falls originated from Mt. St. Helens (Washington) 3,400 years ago, Mt. Meager (Bridge River, BC) 2,400 years ago, and again from Mt. St. Helens 508 years ago. The wind during the 1980 eruption of Mt. St. Helens was to the east. From recent studies it is apparent that, depending on wind direction, the probability of an ash fall at Kelowna from an explosive eruption of one or more of the numerous Cascade volcanoes is quite high for any year.

LAMBLY CREEK VALLEY LAVA FLOWS

The most recent volcanic eruption of a non-explosive nature in the Kelowna area was 762,000 years ago. It originated near Lambly Lake, at the headwaters of Bear Creek on the west side of Lake Okanagan. Several rivers of basaltic lava flowed down Bear Creek, turned south and continued to Lakeview Heights (Figures 5-5, 8-4 and 9-4). The Lambly Creek valley lava flow has been largely eroded by streams and glacial activity but remnants form the cliffs just south of the floating bridge. The vent for the flow has never been found. Chemical similarities between this lava flow and the Chilcotin Plateau lavas suggest that this infrequent and sporadic igneous activity which started 25 million years ago may still be possible today. There is little need to worry however, volcanism only occurs about once every million years, and the last explosive eruption at Kelowna was in the order of 50 million years ago.

HOT SPOTS IN THE OKANAGAN

Brief mention here of geothermal heat resources in the Okanagan serves to emphasize that the ancient volcanic centres have not entirely cooled (Figure 11-10). The Okanagan Valley from Winfield south is believed to have high potential as a geothermal heat source that could be used to heat buildings or generate electricity. Geothermal gradients beneath known volcanic piles range from 50° to 70° Celsius per kilometre of depth (Fairbank and Faulkner, 1992). In other words, a hole 2,000 metres deep could tap water with a temperature of 140° Celsius in some locations. However, recent testing west of Summerland and drilling at the base of Layer Cake Hill in Gallagher's Canyon have as yet not identified an economic energy source. Perhaps higher petroleum prices and hydro-electric costs will be required before the local geothermal heat potential is realized.

Figure 10-10: The white layer shown here is Mazama volcanic ash about one metre thick (note metre stick and back-pack for scale) overlying clay of Lake Oliver sediments (Roed and Fulton, 2011). It was exposed in a borrow pit (Graveyard Pit) 500 metres north of Osoyoos Lake during reconstruction of Highway 97 in 2010. The pit has been re-habilitated and the ash was covered up.

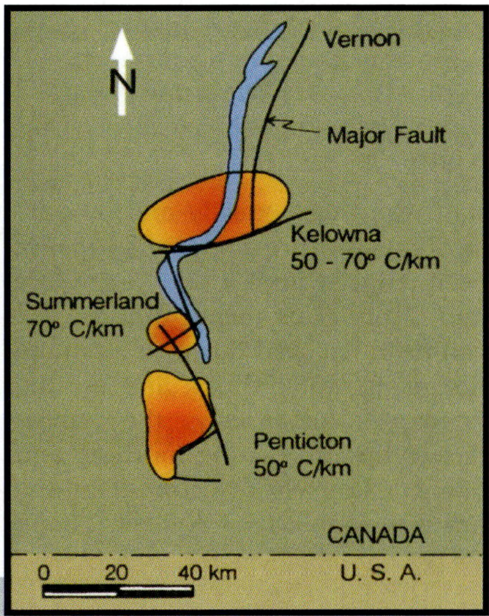

Figure 11-10: Geothermal heat is being generated at certain places in the Okanagan Valley as shown by temperature gradients given in this map (modified after Fairbank and Faulkner, 1992).

GEOHAZARDS OF OKANAGAN MOUNTAIN PARK FIRE OF 2003

By Don Dobson

When moderate to high relief terrain is stripped of vegetation and the binding and protective cover of soil is weakened or destroyed, the terrain and all of its watersheds become vulnerable to a plethora of hazardous geological processes. This is exactly the condition that was created during the Okanagan Mountain Park forest fire that affected 26000 hectares (162 square miles) of watershed terrain adjacent to Okanagan Lake in less than two weeks in the summer of 2003. A firestorm travelling at a speed of 60km/hr swept through several upscale Kelowna neighborhoods like a giant serpent of fire destroying 238 homes.

Figure 13-10: The towering plume of a fire out of control. View to the south at Kelowna.

Figure 12-10 The beginning of the Okanagan Mountain Park Fire south and east of Squally Point.

Figure 14-10: at the edge of the Kelowna Firestorm, Upper Mission area, Kelowna BC, August 22, 2003.

SUMMARY OF THE FIRESTORM

As the summer of 2003 unfolded the weather was spectacular with seemingly carefree endless days of sunshine. All this changed on August 16th. Early in the morning hours a thunderstorm moved up the east side of Okanagan Lake north of Naramata (Figure 12-10). A lightning strike was recorded south of Squally Point in Okanagan Mountain Park. The first smoke from the fire ignited by the strike was reported about 5:30 a.m. The Ministry of Forests Protection staff responded quickly and by mid-morning the fire appeared to have been reduced from a rank 5 fire to a rank 2 fire. This meant that the fire had been knocked down to a ground fire using water drops and ground forces. Unfortunately the winds picked up later in the morning and due to extremely dry conditions the fire was soon out of control [Figure 13-10].

The fire spread rapidly into the park with winds moving it first to the north then to the south toward the communities of Kelowna and Naramata respectively. Fortunately through the efforts of the firefighters and favourable winds the spread of the fire towards Naramata was controlled and although the community was evacuated no homes were lost. Those to the north were not so lucky. The fire at Kelowna claimed its first homes in the Timberline/Rimrock neighbourhood on August 19th. The fire raged into the City in the Okaview-Curlew (Figure 14-10) neighbourhoods on Friday, August 22 then to Crawford Estates (Figure 15-10) by evening. More than 30,000 residents had been forced from their homes, resulting in the largest peacetime evacuation in Canadian history. When Saturday morning dawned the City woke up to find that 238 homes had been lost. The fire then moved back into the forests on the South Slopes burning to the east towards the June Springs neighbour-hood and the KLO Creek and Hydraulic Creek watersheds. By mid-September although the fire spared the June Springs area, 14 of the 18 trestles along the Kettle Valley Railway had been destroyed including all the wooden portions of the large steel structure over KLO Creek.

Figure 15-10: Devastation along Westridge Drive in Crawford Estates, Kelowna. M. Roed's former "house" is circled in red. Photo by D. Dobson, September, 2003.

ASSESSMENT OF GEOHAZARDS

Before the fire was controlled, the City of Kelowna was already busy assessing the damages to property and infrastructure, and evaluating the risks to the City from affected watersheds. It was determined that approximately 15,000 hectares in ten large watersheds draining through the city had been either partially or completely burned. The hazards facing the city were determined to be related to changes in hydrology leading to potential erosion, flash flooding, debris flows, and other processes of slope instability. Of particular concern was the risk from flash floods if intense fall rainstorms occurred within the burned watersheds because the risk may be significantly increased if hydrophobic (water-repellent) soils were extensive in the watersheds. Hydrophobic soils are a by-product of the burning of the organic material on the ground surface as well as the organic matter within the upper layer of soil by a very hot wildfire. Some of the gases created by the burning material condense within the soil forming a waxy substance that binds the soil particles together into a waterproof layer.

On October 22nd, 2003 an intense rainstorm did occur over the watersheds from Bertram Creek in the west to Rembler Creek in the east. There was severe erosion and flooding in these watersheds with peak flows estimated after the event ranging from six to ten times the 200-year flood flows calculated for the streams under prefire conditions (Figure 16-10). These flows were the direct result of extensive hydrophobic soils in the burned watersheds. There was considerable damage to city roads as well as to a number of dwellings but fortunately no injuries or lost lives.

The persistence of the hydrophobic soil condition is not well understood but it is expected to disappear over a three to five year period. Based on engineering assessments that included the risk of floods while water repellent soils remained, the City implemented a program to upgrade channels and road crossings at a cost of approximately two million dollars. Most of this upgrading was completed early in 2004 before the spring freshet.

Figure 16-10: Severe erosion due to rainstorm on October 22, 2003 in the Rembler Creek Valley. This erosion is judged to be directly related to deforestation from the Kelowna Firestorm. Extreme climatic events in the next five years or so, may have similar consequences. Photo by G.VanEmmerik, October 23, 2003.

COMPARISON TO EARLIER FOREST FIRES

The Okanagan Mountain Park Fire was not the first time the forests on the South Slopes have burned, but it is the first time that a wildfire became what is referred to as an interface fire.

An interface fire is a wildfire in the forest that burns into an adjacent urban development. The forests in the valley bottom and lower portions of the South Slopes are referred to as the Interior Douglas Fir biogeoclimatic zone forests. The fire history in these stands involved frequent small ground fires known as under burns that reduced the ground fuels and maintained the stands. Then there are infrequent major fires known as stand replacing fires that actually burned the stands. This is what the Okanagan Mountain Park Fire was. Such major fires have a frequency that is measured in the hundreds of years.

Based on the age of the stands in this area, there had not been a major fire, of the significance of the Okanagan Mountain Park Fire, for more than 100 years. At higher elevations in what is known as the Montane Spruce biogeoclimatic zone, stand-replacing fires are frequent, typically every 100 to 150 years. The Ministry of Forests has maps showing the locations and extent of fires that burned through the upper KLO and Bellevue Creek watersheds in the 1930's. There are still some of the older residents in the valley who remember these fires and the impacts on the watersheds. It was recalled by Mr. Browne-Clayton, before he passed away in September 2003, that Bellevue Creek would turn Okanagan Lake black each spring for years after the fire.

With the expansion of the urban development into the forested areas in the valley, the risk of future interface fires will continue to increase. The provincial government and local governments are working on plans to reduce the fire risks in these areas but it is not known what the costs will be to implement such plans. What we do know is there will be more wildfires in the valley that will place resident's lives and property at risk. Like it or not, fire is a natural force in our forests, always has been and always will be.

Figure 17-10: BX Falls, Vernon, BC, acrylic painting by M.Roed.
BX Falls is located off Tillicum Road, on the way to Silver Star Mountain near Vernon, BC. The waterfall is along BX Creek that here has incised hard granitic rocks of the Monashee metamorphic complex.

CHAPTER 11

GEOTECHNICAL CONDITIONS IN THE KELOWNA AREA

by Norm Williams and Murray A. Roed

The term geotechnical refers to the study and prediction of how bedrock, surficial deposits, groundwater, surface water and topographic conditions relate to construction of buildings, roads and utilities. Even a casual glance at the surficial geology map of the Kelowna area (Figure 4-5) suggests that the range of deposits presents demanding challenges to any geotechnical consultant advising on the use of the land. Conditions vary considerably and are briefly summarized.

DELTAIC AND ALLUVIAL FANS

Extending roughly from Orchard Park to the lakefront Kelowna is situated on a thick sequence of interbedded organic material, clay, silt, sand and gravel deposited by ancestor channels, ponds and fans of Mill (Kelowna) Creek and Mission Creek. These sediments are loosely consolidated and are usually saturated within a few metres of the surface. The finer grained soils are loose and compressible and even for small buildings require special consideration for settlement and earthquake behavior.

GLACIAL LAKE SEDIMENTS

Silt and clay deposited in Glacial Lake Penticton are widespread in the area. These deposits can be found in Glenmore Valley, at Okanagan University College, along the base of Knox Mountain, along Swamp Road and Hartman Road, and in places in East Kelowna and KLO Road as well as numerous locations on the west side of the lake. These soils are hard clayey type sediments with some swelling potential. That is, when wet they swell, and when dried they shrink. Numerous incidents of settlement, cracking of foundations and basement slab displacements can be traced to the behavior of these soils under differing conditions.

Generally, for single family dwellings, these clayey soils should be removed to a certain depth and replaced with compacted granular materials that have good drainage. In some cases an alternate design should be considered.

GLACIAL MORAINE DEPOSITS

Moraine deposits include till (Figure 7-9), a dense mixture of clay, silt, sand and boulders, deposited directly from melting glacier ice, and stratified sand and gravel deposited by glacial meltwater. These surficial soils are most common at elevations above 500 metres. Construction and foundation problems are few since they are usually well drained, have high strength and are not compressible under normal loads. However, where drainage is poor and where slopes are involved these soils can be hazardous (see Chapter Ten).

BEDROCK

A great variety of bedrock types occur in the Kelowna area as shown in Figure 1-4. Generally, bedrock does not offer many geotechnical problems unless the rock is fractured or loose, or if it is subject to rockfall, or some other form of failure. In many rock localities the greatest impact is economic in that the rock may have to be blasted which increases the cost of construction and introduces a potential hazard if blasts are not properly controlled.

GEOTECHNICAL IMPLICATIONS OF THE FLOATING BRIDGE

Opening of the Okanagan Floating Bridge by Princess Margaret in 1958 ended a ferry service that dated to 1904. Over 30,000 people attended the opening which also marked the first live televised event of CHBC-TV (Evans 1993). The bridge cost 7.5 million dollars to construct. Now, in 2004 a proposed bridge expansion is expected to cost about 100 million dollars and 20 to 30 million for approach improvements (Ron Westside, P.Eng, personal communication). Along with its social, economic and political ramifications, it is a major geotechnical and structural engineering achievement (Pegusch, 1957). The bridge is over three kilometres long, including approaches. It has a 640 metre long floating section consisting of twelve reinforced concrete pontoons rigidly connected to form one continuous pontoon. The pontoons are held in place by cables anchored seven metres into lake bottom sediments. One of the original concrete anchors, weighing 64 tonnes,

is displayed at the corner of Harvey Avenue and Ellis Street in downtown Kelowna (Figure 1-11). In the eastern section of the bridge a 79 metre lift span accommodates the passage of large boats, such as barges that are no longer in use on the lake.

Why build a floating bridge? A suspension bridge was designed originally but the highly compressible(soft) surficial sediments in the lake bottom ruled it out because there was no suitable layer in which piling could be practically and economically installed. There would also be a danger of liquefaction (loss of strength due to shaking) in the event of an earthquake. These deltaic and lacustrine sediments include soft layers, and are completely saturated. Bottom materials could not support a rock causeway either, which was also a design option. Thus geological conditions and economics necessitated a floating bridge. Additional crossing capability is now required, and all of the options are being re-evaluated including an update on earthquake hazards. A new bridge is slated to be completed by 2008 (see Addendum).

Figure 1-11: One of the anchors originally used for the Kelowna floating bridge is displayed in Anchor Park at the corner of Ellis Street and Harvey Avenue. Photo by Debra Roed.

THE GRAND OKANAGAN LAKE FRONT RESORT

Clay, silt and sand associated with deltaic and lacustrine sedimentation also underlie a twelve storey hotel at the Grand Okanagan Lake Front Resort in down town Kelowna (Figure 2-11).

Elaborate remedial measures were necessary to prepare this site for construction (Howie et al, 1994). Soils to depths of 55 metres were investigated by drilling and engineering tests. They were found to be compressible at depth, and an upper sand layer containing high silt content had potential for liquefaction.

The entire site was loaded with a ten metre high pile of soil to squeeze out water and force settlement of soft layers in the subsurface (Figure 2-11). This load was maintained for about one year in 1991 and served to consolidate the deep soils to an acceptable engineering standard. Vibro-compaction was then used to densify the upper sands at the site. Finally a "raft" slab of concrete 1.2 metres thick was placed to form the foundation of the structure and further limit long term settlement of the subsoil. A six metre wide strip along the lakeshore was also densified to arrest potential lakeshore instability.

Geotechnical design for this development has resulted in settlement below predictions. It has prepared the soil to withstand the risk of an estimated 1 in 475 year shaking event due to an earthquake, and allowed the possibility of further development.

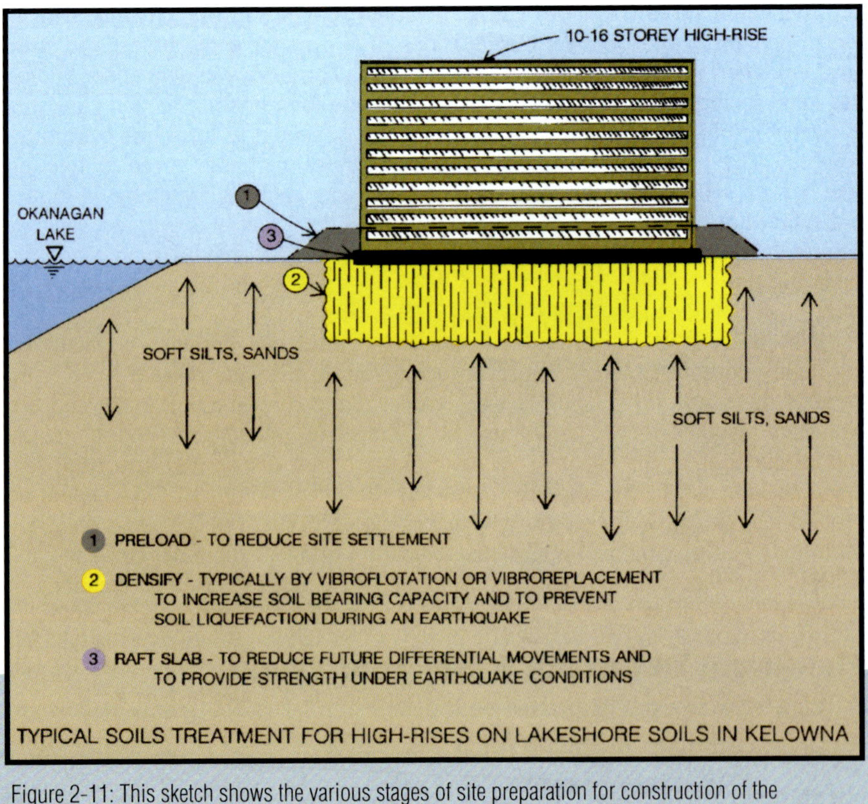

Figure 2-11: This sketch shows the various stages of site preparation for construction of the Grand Okanagan Hotel beside the lake in downtown Kelowna.

Historically the site of the Grand development first served as a ferry landing. Between 1900 and 1930 part of the site was used for waste disposal from a sawmill. Since then the shoreline has been continuously pushed into the lake by random filling and finally by a dyke built in 1986 and subsequently infilled.

UNIVERSITY OF BRITISH COLUMBIA OKANAGAN

During excavation for foundations at the University of British Columbia Okanagan (UBCO) formerly Okanagan University College, across from the airport, clay subject to high swelling was encountered (Figure 3-11). Final design was based on removal of over three metres of the clay and replacement with compacted sand and gravel. This acts as a thick "mat" foundation, and will insure that no settlement takes place even if there is minor swelling in the remaining clay soils at depth.

OTHER CONDITIONS AT KELOWNA

Clay, silt and sand flat terrain of Kelowna is generally poorly drained. A high water table is characteristic and basement flooding is a common problem in many localities. Along Swamp Road and adjacent terrain, clayey soils have a high organic content which makes them subject to settlement and swelling. Consequently the road was warping. Construction of a stable road base in organic soils may require excavation and replacement with granular material or placement of fill with adequate drainage. Either way costs escalate. The road was rebuilt in 2005.

Unusual soils occur in places. Near the Bernard Avenue overpass by the Apple Building a 60 centimetre thick layer of diatomaceous earth was encountered in flood plain sediments. This probably represents deposition in an old pond related to a meander of one of the streams that used to cross the area.

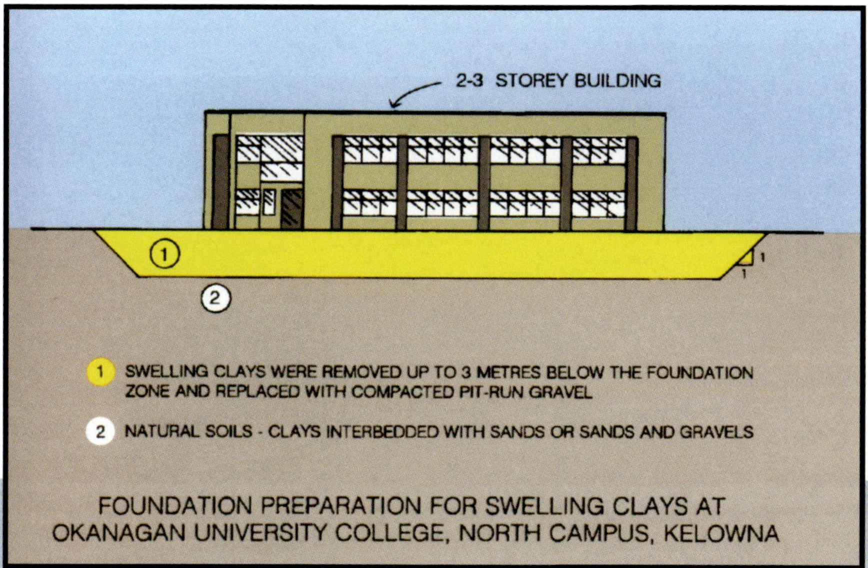

Figure 3-11: Foundation preparation for the north campus of the UBC Okanagan is summarized in this sketch.

Figure 4-11: Manhattan Point, Lake Okanagan, acrylic painting by M.Roed. About the time Father Pandosy arrived at Kelowna, or Okanagan Mission, the beach at Manhattan Point might have looked like this. The view to the south shows the rugged granitic hills of what is now Okanagan Mountain Park, through which Father Pandosy made his way. The hill in the middle is Mount Boucherie, once a mighty volcanic dome, greatly reduced by multiple glaciations. The south end of Hayman (McDougall) Ridge is visible on the right.

CHAPTER 12

MINERAL RESOURCES

by John D. Greenough, Brian Hughes, Dave Capstick and Murray A. Roed

---

Mineral resources have played an important role in the Okanagan's history. Just two years after Father Pandosy established his agricultural based Okanagan Mission in 1859, gold was discovered in Mission Creek, fueling early settlement in the valley. A silver mine at Beaverdell helped bring the Kettle Valley Railway into the area in the early 1900's and development of a molybdenum-copper deposit by Brenda Mines Ltd. poured millions of dollars into the local economy during the 1970's and 80's. A uranium deposit discovered at Lassie Lake south of Big White Mountain in 1977 did not bring more wealth into the local economy. However, it precipitated a moratorium on uranium mining and exploration in British Columbia which reflects the environmentally conscious attitudes of area residents a decade and a half ago. These four ore deposits represent the most important non-renewable mineral resource discoveries affecting Kelowna, apart from important sand and gravel aggregate resources. The geologic formation of these mineral resources, discovery and development are explored below.

GOLD MINING, GALLAGHER'S CANYON

by Murray A. Roed

When Father Pandosy founded the Okanagan Mission in 1859 on the floodplain of Mission Creek, the settlement was primarily an agricultural production centre, independent of fur trading or gold mining which were the historic economic targets of early explorers. According to historian Primrose Upton, gold was first discovered in the Okanagan as early as 1833. The discovery, by botanist David Douglas, apparently came as a result of panning on the shores of Lake Okanagan.

Gold in Mission Creek valley at what is now known as Scenic Canyon Regional Park was discovered by one of Father Pandosy's original guides, William Peon (also spelled Pion), in 1861, the same year that the rich deposits at Williams Creek and Lightning Creek in the Cariboo were discovered, and long before the Klondike gold rush of the 1890's. Placer mining continued at Mission Creek until about 1900 but work was sporadic and probably few made more than wages. Chinese miners also worked the stream valley for a period of time and left visible evidence of their activity. They always piled the rocks from their diggings in neat fence-like structures. Dr. George M. Dawson, one of the most prominent scientists of the Geological Survey of Canada, and a pioneer in the annals of Canadian geology, visited the mine workings on Mission Creek in 1877. He reported only a few men still employed. Early miners apparently recovered two to three ounces of gold per day from the stream bed, and the gold was quite coarse.

Ed Aldredge, a Penticton historian, places the discovery of gold on Mission Creek at least as early as 1853. He also mentions that the late Alan Richardson of Penticton had a claim on Mission Creek, somewhere, as recently as the late 1970's.

Dan Gallagher was probably the most persistent miner. According to historians Lundy and Zoellner (1990), he pre-empted land in the canyon in 1898 and worked a claim up to the 1930's. He lived there as a recluse in a cabin, was in popular demand as a fiddler, and is rumored to have been a cattle rustler on occasion. His cabin burned, and its location is not known. Today the locality is known as Gallagher's Canyon which apparently includes the total length of rugged bedrock landscapes along Mission Creek.

THE BRENDA MINE

by John D. Greenough

The Brenda Mine, about 60 kilometres southwest of Kelowna (Figure 1-12), has an interesting exploration, development and decommissioning history that underscores the perseverance and ingenuity of the Canadian mining industry. In the 1930's the Sandberg family of Kelowna discovered copper and molybdenum minerals at the location that ultimately became the Brenda open-pit mine. Bob Bechtel of Penticton re-discovered and staked the long-abandoned Sandberg workings in 1954, and optioned the property to Noranda following discussions with Bern Brynelson. Drilling in 1956 indicated an average ore grade of 0.21% copper and 0.03% molybdenum. These results enticed Kennecott Copper to join Noranda and carry out a systematic surface exploration program in 1957. Discouraging results saw the property returned to Bechtel with sufficient assessment work done to keep it in good standing for 10 years.

By 1964 the development of large-tonnage open pit mining methods in other places, and increasing demand for molybdenum from the steel industry, prompted Bern Brynelson, Morris Menzies and Mervin Davis to form a publicly traded company (Brenda Mines Ltd.) on the Vancouver Stock Exchange to carry out further exploration. Brenda Mines Ltd. in partnership with Nippon Mining of Japan undertook a $3.5 million mining feasibility program in 1965. This involved collecting a 10,800 tonne bulk sample of the mineralized zone. The bulk sample was run through a 90 tonne per day mill to determine what percentage of the molybdenum and copper could be recovered from the ore.

Enthused by the results, Noranda bought a controlling interest in Brenda Mines Ltd. and financed construction of a $62.5 million mining operation in January of 1968. The ore body had proven reserves of 159.3 million tonnes grading 0.183% copper and 0.049% molybdenum, making it the lowest grade Cu-Mo mine in the world! Thus, following 16 years of perseverance by Bechtel and as a result of the ingenuity of Noranda geologists and engineers, the Brenda ore body was mined profitably for the next 21 years.

GEOLOGY

The Brenda deposit occurs at the junction of two large igneous intrusions, the older is a diorite and the other a granodiorite (Soregaroli and Whitford, 1974). Both are part of a batholith that formed some 146 to 176 million years ago during the Jurassic Period (Figure 2-12). Similar rocks of this age are quite common in the Interior Mountain System Plateau (Peto, 1973). Molten rock forming the batholith was intruded (injected) into late Paleozoic to late Triassic (200 - 250 million years old) volcanic and sedimentary rocks. The intruded rocks have been "baked" or metamorphosed for a distance of 500 metres away from the granite contact. Detailed geologic mapping indicates that the Brenda ore deposit formed at the roof of the batholith, near where it contacted older volcanic and sedimentary rocks (Rice, 1947). The ore body is cut by various other igneous rocks, many probably intruded during the Tertiary mountain building event. A period of prolonged uplift and erosion, including glaciation, following the Tertiary generated the deeply dissected valleys, gently rolling upland plateau and present-day bedrock outcrop pattern (Figure 3-12).

The ore body occupied an irregular area of approximately 900 by 400 metres and extends downward more than 300 metres. The main ore minerals are chalcopyrite (a copper, iron and sulphur mineral) and molybdenite (contains molybdenum and sulphur). These minerals, which represent less than 1% of the rock, are accompanied by variable amounts of pyrite (fool's gold), minor bornite (another copper mineral) and rarely sphalerite or

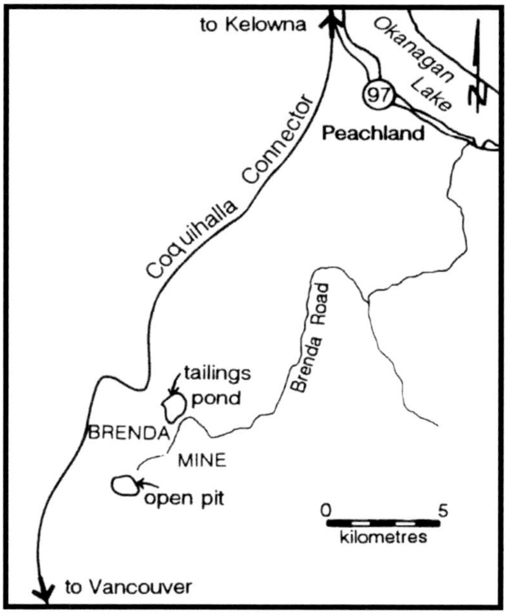

Figure 1-12: Location of the Brenda Mine.

galena (zinc and lead ore minerals respectively) together with the waste (gangue) minerals quartz, potassium feldspar, biotite, calcite and epidote.

The ore mineralization occurs in fracture-filling veins, typically 0.5 to 2.5 cm thick with the number and size of veins, and thus ore grade, increasing toward the centre of the deposit. At least five distinct stages of veining are recognized based on mineralogy of the veins.

MINERAL GENESIS

The Brenda deposit belongs to a class of ore deposits known as Copper-Molybdenum porphyries. These deposits form when a continent collides with ocean crust resulting in the ocean floor being shoved beneath the continent. This plate tectonic process is called subduction. It causes rocks at the bottom of the continental crust to become so hot that they melt. The hot molten rock, or magma, is less dense than surrounding rocks and thus more buoyant. The magma forcefully rises up through the crust as a large (kilometres wide by many kilometres deep) "blob". In some cases it reaches the surface forming a volcanic eruption. More commonly the magma does not reach the surface, and an intrusion forms because cool rocks near the surface of the earth cause outer portions of the blob of magma to solidify (crystallize). While the magma rises through the crust, slow cooling causes large crystals to grow but on reaching cooler rocks the outer portion of the intrusion cools quickly forming many smaller crystals surrounding the larger ones. The resulting rock contains two grain sizes and is known as a porphyry. At this point an ore deposit has yet to form.

After the intrusion stops moving, its interior portions are still molten and many thousands of years are required for it to crystallize. Water present in this magma becomes concentrated because minerals forming from the magma do not have water in their structure. These

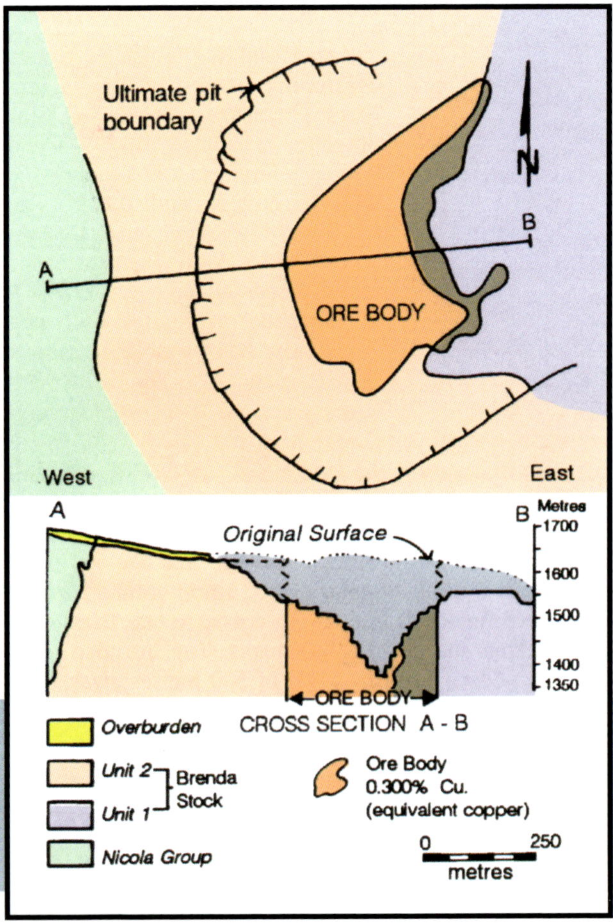

Figure 2-12: Geology of the Brenda Mine (after Soregaroli and Whitford, 1974).

hot-water fluids rise from depths measured in kilometres through the crystallizing magma and become trapped below the completely crystallized roof of the intrusion near the surface. As they rise they pick up elements such as copper, molybdenum, iron and sulphur that, like the water, are not used by minerals growing from the magma. As these fluids move upward there is less pressure on them and they begin to expand. The expanding fluids eventually become so concentrated near the top of the intrusion that they shatter the cool solidified outer rim of the intrusion. The hot fluids go rushing through the fractures but as temperature and pressure drop, dissolved elements begin to precipitate. At Brenda chalcopyrite and molybdenite, along with the gangue minerals mentioned previously, were precipitated from the fluids thus filling the fractures and forming the ore body.

PRODUCTION HISTORY

Low grade mineral deposits require highly mechanized, bulk mining practices that process large amounts of ore daily at low cost per tonne. The Brenda Mine was designed for a throughput of 21,800 tonnes per day, however, the innovative and industry-first use of computer-controlled equipment and methods allowed the mine to substantially raise this production level. By 1989 a daily throughput of over 30,000 tonnes was achieved at an operating cost of just $4.39 per tonne milled!

Mining operations came to an end in April 1990, as a result of an eight million tonne rock slide on the west wall of the Brenda Pit (see Figure 6-10). By that time 182 million tonnes of ore had been milled and 109 million tonnes of waste rock moved. Concentrates of copper minerals were shipped to Japan whereas molybdenite concentrate went to Belgium where 271,938,021 kg. of copper, 65,469, 525 kg. of molybdenum, 122,814 kg. of silver and 1,777 kg. of gold were smelted (extracted) over the life of the mine. The economic benefits during the Brenda production period were $449 million in the purchase of supplies, $215 million in wages and benefits, $79 million paid in taxes and $22 million distributed as dividends to the shareholders of the company (Harris, 1992).

MINE DECOMMISSION

With mining operations finished, Noranda Mining and Exploration Inc., the owners of the Brenda Mines site have been committed to returning the land and water to a suitable post-mining condition. Buildings have been removed, soil re-established and vegetation suitable to the local environment nurtured through aerial seeding, fertilization and irrigation of newly planted areas. Between 1990 and 1997 a total of $38 million dollars was spent on closure and reclamation.

After the mine ceased production, Noranda was faced with new technological challenges. Precipitation falling on the mine site was collected in either the tailings pond or the open pit. This was necessary because as water flows over exposed rocks on the site, it picks up various metals but most importantly molybdenum. High concentrations of molybdenum meant that the water could not be directly released into the environment. In order to diminish the impact of reduced amounts of water entering local streams, and to minimize the amount of water requiring treatment, a fresh water diversion system was created to keep precipitation (clean water) from coming into contact with waste rocks containing elevated molybdenum concentrations.

Figure 3-12: Westerly view of the Brenda Mine site showing the tailings pond and dam (to right) in the foreground and mined-out pit in the background. Photograph by John D. Greenough piloted by Craig Merkl, Spring 2004.

A water treatment facility costing $11 million dollars was constructed and went into operation in 1998 to treat contaminated water collected and stored on the site. The treatment process involves adding sulphuric acid to lower the pH (increase the acidity) of the water. Then, Fe-bearing chemicals are added that form precipitates (solid iron-molybdenum particles) that bind with molybdenum and remove it from the water. These solids either settle forming a sludge or are filtered out. The sludge is separated from the treated water and stored in specially designed cells. The resulting water, ready for discharge, contains on average 0.039 g of molybdenum per million g of water, much less than the permit limit of 0.25 g of molybdenum per million grams of water (< 0.25 mg/L). The permit authorizing the treatment and discharge of water into the Trepanier Creek watershed also stipulates that down-stream at Peachland the maximum average concentration of molybdenum is only 0.03 – 0.06 mg/L depending on the time of year. This more than meets national and international drinking water standards. Discharge from the treatment facility is expected to average about 3 million cubic metres per year at an operating cost of $1.0 - $1.5 million. Brenda is still contributing to the local economy!

HIGHLAND BELL MINE

by Brian Hughes

The abandoned Highland Bell Mine is located 1.6 km east of Beaverdell (Figure 4-12) on the western slope of Mount Wallace and was nationally famous for its silver production.

HISTORY

Work on the mineralized veins in the Beaverdell area dates to 1889, and records indicate intermittent silver production from 1900 to 1913. Production became continuous in 1913 with the arrival of the Kettle Valley Railway.

Four separate mines called the Bell, Highland Lass, Beaver, and Sally (Figure 4-12) produced ore until 1946 when Lietch Gold Mines Ltd. obtained control over all production under the name of Highland Bell Beaverdell Mine. Ore was sent by rail to a smelter at Trail until a 50 tonne a day concentrator was built in 1950. Afterward only concentrates were sent to Trail. Teck Corporation acquired the mine in 1970 and continued production until the mine was closed in February of 1991.

GEOLOGY

The main economic vein system of the Beaverdell mine occurs in a set of fractures in the West Kettle quartz diorite to granodiorite, part of the Nelson Batholith of Jurassic age. Thus the ore-forming processes, which involved movement of hot water solutions through a fracture system, resemble those at the Brenda Mine except that the metals in the solutions were different. Metamorphic rocks of Paleozoic and Early Mesozoic age occur as roof pendants (pieces of country rock that hang down into the granodiorite). Porphyritic quartz monzonite of the Valhalla intrusions of Jurassic to Cretaceous age intrude the Nelson batholith rocks west of the mine.

MINERALIZATION

Veins of quartz with smaller amounts of calcite and some fluorite formed along easterly to northeasterly trending faults and fractures. These veins also contain pyrite, galena and sphalerite, and lesser amounts of pyrrhotite, chalcopyrite, arsenopyrite, polybasite, argentite, and locally, close to fault zones, native silver wires and plates. Silver is usually associated with galena, sphalerite and antimony minerals and gold with pyrite and chalcopyrite.

PRODUCTION

Due to extensive block faulting of the deposit, accurate reserves have been difficult to predict. The mill was expanded to handle 108 tonnes per day in 1954 and the mine continued at that production level until its closure. A total of 33,500 tonnes of ore were mined in a typical year (1979) and yielded 388 tonnes of lead concentrate, 399 tonnes of zinc concentrate and 97 tonnes of silver from a jig concentrate. In 1979 gross metal content amounted to 4199 g of gold, 10,259,637 g of silver, 613 kg of copper, 93,324 kg of lead, 140,679 kg of zinc and 1000 kg of cadmium. Production over the life of the mine exceeded one billion grams (300 million ounces) of silver, twelve million kilograms of lead, fourteen million kilograms of zinc and significant quantities of gold, cadmium and copper.

URANIUM DEPOSITS

by Brian Hughes

During the late 1960's interest in uranium exploration was sparked by the metal's price rise to $40.00/lb and the increasing dependency of many nations on nuclear power. Of particular importance to the Kelowna area were the efforts of Power Reactor and Nuclear Fuel Development Corporation (PNC), a Japanese company. They began a search in the Kelowna area for basal uranium deposits similar to those being mined in the Ningyo Toge and Tono deposits in Japan. The initial discovery of basal type uranium deposits was made at the Fuki outcrop northeast of Beaverdell (Figure 5-12) during a car-borne regional scintillometre survey. A scintillometre is a device that measures radioactivity given off by uranium-bearing minerals. Continued scintillometre surveys, geochemistry studies of soils, stream silts and groundwater, regional mapping and widespread drilling led to the discovery of another deposit by PNC geologists, the Tyee just north of Hydraulic Lakes. As other exploration groups became involved deposits were found at Haynes Lake and, the largest deposit, the Blizzard property. The Blizzard deposit has reported reserves of over 2.1 million tonnes grading 0.226% uranium. However, in 1970 the British Columbia government imposed a moratorium on all uranium exploration and mining and to date all of these deposits remain undeveloped.

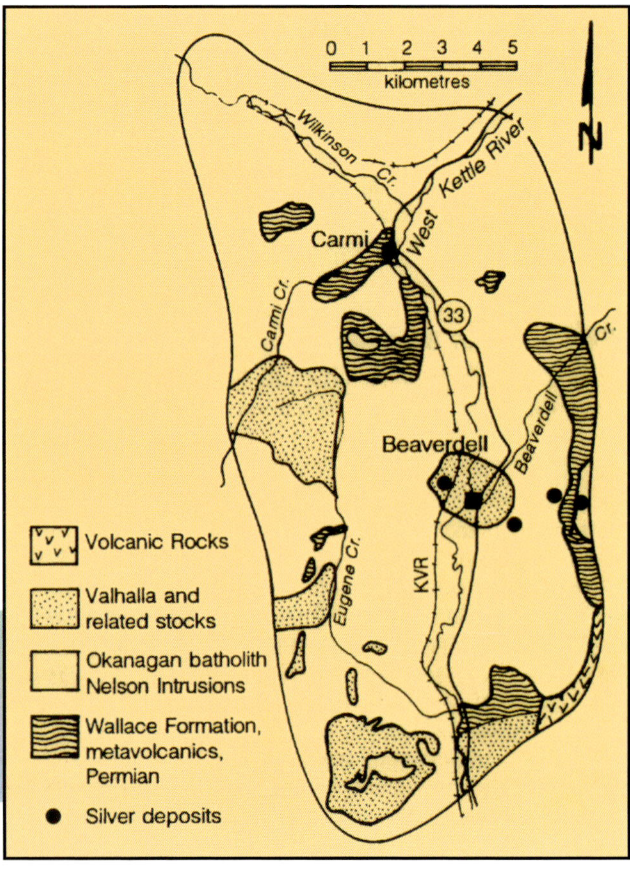

Figure 4-12: Geology of the Beaverdell area and location of silver deposits. Beaverdell is 80 kilometres east of Kelowna along Highway 33. Map modified from Godwin et al, 1986.

REGIONAL GEOLOGY

The regional geologic setting in this portion of the Okanagan Highland consists of the uplifted Shuswap metamorphic rocks intruded by the Valhalla granitic rocks of Mesozoic age (Figure 4-12). Further intrusion took place during the Tertiary to form the Coryell intrusions composed of monzonite and syenite. During the Pliocene large north-south trending streams developed on the surface of these rocks and widespread sand and gravel deposits with carbonaceous material (organic remains of plants) were laid down in the stream channels. These sediments were later overlain by plateau basalts of late Tertiary age (see Chapters Two, Three and Four) which flowed down into the valleys.

ORIGIN OF THE URANIUM DEPOSITS

The mineralization was likely derived from the Valhalla and Coryell intrusive rocks that are known to have high background Uranium contents. Uranium mineralization was deposited in the more carbonaceous parts of the river sediments that were covered by basaltic lava flows (Figure 6-12). It is believed that groundwater percolating through the uranium-bearing intrusive rocks dissolved the uranium and carried it downward. When the flow path of these solutions intersected carbonaceous parts of the buried stream sediments, reducing (oxygen-poor) conditions were encountered which caused the precipitation of the uranium minerals. Mineralization varies between deposits and includes uranium oxides and phosphates such as saleeite, ningyoite and autinite. Other minerals associated with these deposits are iron sulfides and oxides such as pyrite, marcasite and hematite. Most mineralization is observed as coatings on sediment grains and as a partial cement in the matrix of the sand and gravel sediments.

Continued uplift of the Okanagan Highland terminated this uranium enrichment process. Where later glacial erosion was not too deep, these basal type uranium deposits were preserved below protective caps of plateau basalts.

A typical basal type environment of stream troughs in weathered pre-Tertiary rocks filled with clastic sediments and overlain by plateau basaltic rocks can be seen in cross-section in several road outcrops in the area. One of the best sections can be viewed on the Okanagan Falls logging road approximately 2.5 kilometres south of its junction with the Big White access road east of Highway 33 (Figure 5-12). At this location several basaltic flows displaying columnar structure (referred to as the West Kettle Columns) are exposed and overlie sand and gravel beds of Tertiary age which in turn overlie weathered and leached pre-Tertiary intrusive rocks. Also, several sections of the same type can be viewed on the Kettle Valley Railway between KLO Creek and Bellevue Creek trestles. The only section with significant uranium mineralization that can be seen is the Fuki showing near Deer Creek located at Kilometre 22.8 on the Beaver Creek forestry road east of Beaverdell.

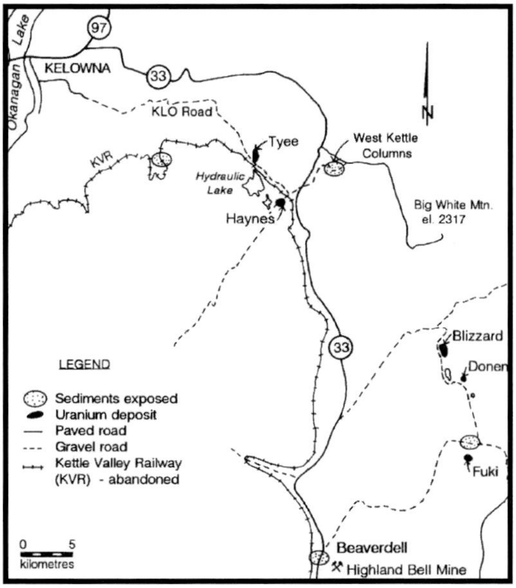

Figure 5-12: Uranium deposits near Kelowna.

REPORTED ORE RESERVES

Drilling has indicated that the uranium deposits are commonly channel shaped varying in width from less than 10 metres to over 100 metres, and ranging from less than a metre thick to over 40 metres in the case of the higher grade Blizzard deposit.

SIGNIFICANCE OF THE URANIUM DEPOSITS

The discovery and development potential of these uranium deposits triggered strong environmental opposition in the Okanagan region. This opposition forced the British Columbia government in 1970 to retroactively terminate all uranium exploration in the province. Companies have never been compensated for their exploration costs.

Apart from demonstrating the effectiveness of environmental lobbyists and in contrast the ineffectiveness of the exploration industry to present defensible counter-arguments, there is an economic significance to this. The total gross value of uranium ore in the deposits exceeds 125 million U.S. dollars in 1995. The loss of this has to be weighed against other cost-benefit factors, some of global importance such as the possible abuse of radioactive materials. It also has to be measured against the successful uranium mining industry that has helped sustain the provinces of Saskatchewan and Ontario for many years. One fact is sure, it sends the message out to the world's mineral industry that British Columbia can be a tough place to do business.

Deposit	Tonnes of Ore	Grade	Tonnes of U
Blissard	2.2 million	0.183%	4020
Tyee	2.1 million	0.031%	J650
Conen	1.0 million	0.040%	J400
Fuki	0.5 million	0.025%	J125
Haynes Lake	2.0 million	0.017%	J340
North Tyee	1.0 million	0.017%	J170

AGGREGATE

Aggregate refers to sand, gravel and rock deposits used for construction purposes. The majority is used for the manufacture of concrete and pavement. It is a substantial industry which accounts for 5% of the total mineral production of British Columbia.

Most of the sand and gravel deposits that are presently developed in the Okanagan consist of glaciofluvial deposits formed when the Fraser glacier melted. A major meltwater channel is located on the west side of the lake, another in the vicinity of the airport, and several along the East Kelowna Escarpment. Parts of these deposits have been mined for sand and gravel. Many extensions of this valuable resource now lie beneath housing developments (for example, Crawford Estates in the Mission) or are in the Agricultural Land Reserve. For aggregate producers in the Okanagan, there are no other nearby deposits. The industry locally is facing a crisis, and the public will be forced to pay increased costs as deposits further away are utilized. This is a province-wide phenomenon that is just now receiving a little attention.

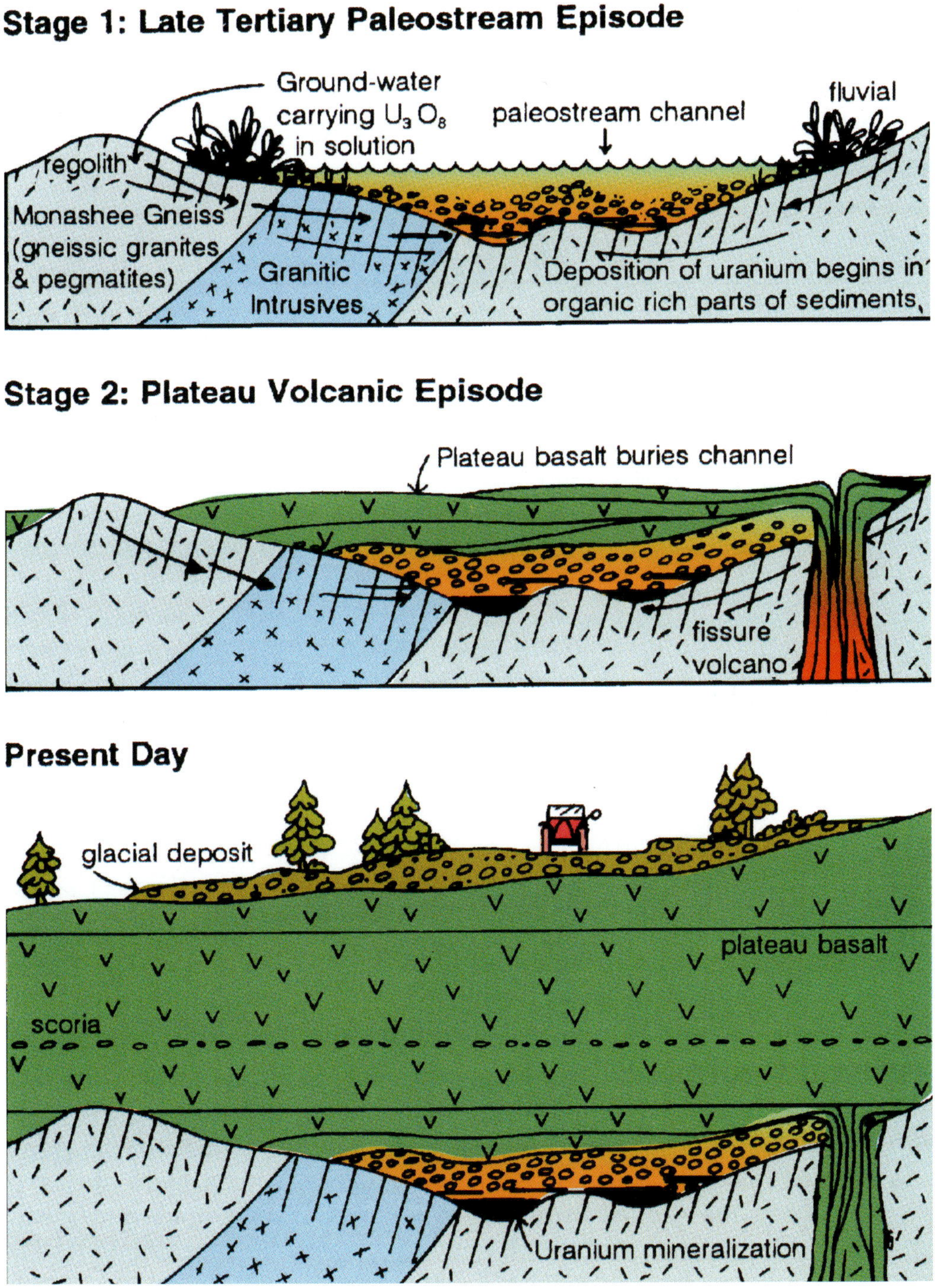

Figure 6-12: A sketch illustrating various stages in the formation of the uranium deposits near Kelowna.

GEMS IN THE HILLS

by Dave Capstick

Volcanic rocks of Tertiary age in the Kelowna area harbour a variety of mineral types valued by collectors, lapidary specialists and rockhounds. The minerals form in cavities and fractures in lava flows where mineralized fluids collect and precipitate silicate minerals. Broadly the minerals consist of two types, those where crystals are visible, and those that are cryptocrystalline (crystals not visible). Crystalline varieties include crystalline quartz and rarely amethyst quartz. Cryptocrystalline varieties include chalcedony, agate, variously coloured chert , jasper and opal. Potential collecting zones are shown in Figure 1-9 – and some examples of cut and polished minerals from the Kelowna area are shown in Figure 7-12.

The gems occur in three forms, as fracture fillings, as partial cavity fillings and as solid egg-like nodules. Any exposed rock face may display fracture fillings and cross-sections of cavity fillings, however the best place to look is along talus slopes where rocks have spalled off and collected at the base of rock ledges. Certain formations in the volcanic rocks are more favourable hunting ground. One such unit occurs in the Hayman Lake upland area and along the western side of Bear Creek valley. Here, inconspicuous tan coloured pebbles can be found scattered on the surface poking out of moss or covered with lichen in lightly forested terrain. Cut them open to reveal their internal structure. The solidly filled ones are sometimes referred to as "thunder eggs". Others that are partly filled and have a druzy lining of crystals are called geodes.

A range of collector minerals occur in the light coloured volcanic rocks of Mount Boucherie. The most common is white to light gray banded translucent agate with a hint of opal in places. Fracture fillings and cavity linings of well formed quartz crystals are common, and amethyst quartz has been reported.

The slopes adjacent to Rose Valley reservoir contain banded agate and chalcedony to the east, and druzy quartz and colored chert in veins to the west. The volcanic rocks here also display a prominent cave and arch structure. The eastern shoulder of Grouse Mountain has agate seams, large geodes; green opal and seams of chert occur along the northwestern slope. All of the terrain adjacent to Mount Swite, Mount Law, Carrot Mountain and Upper Powers Creek contain agate, chert veins, geodes and some opal. Green olivine (peridot) has been found in basalt volcanics capping Carrot Mountain. North of Black Knight Mountain chert veins, small agates and some opal occurs associated with contorted columnar volcanics.

Fossils are largely restricted to the White Lake Formation. Most of the fossils are plant fragments associated with coal layers, but there should be fish, reptile and amphibian fossils in this formation as well as petrified wood. Good examples of the coal and plant remains can be found on the eastern side of the highway near the Glenrosa interchange. In places, entire parts of tree trunks, turned to coal, occur. Poorly preserved fossils occur in marble between Bear Creek Park and Traders Cove and near Paul's Tomb (#34, Figure 1-9). Productellid brachiopods occur in the marble hills north of Okanagan Landing in Vernon.

Okanagan Opal was discovered northwest of Vernon in 1991 by the Yorke-Hardy family. Since then a unique industry has emerged to market these precious gems. Fiery red, green yellow and blue opal occurs in fractures as thin veins in tuffaceous volcanic rock (a rock formed as volcanic dust settles). Most of the veins are 1 to 3mm thick but rarely a 10mm thick lense is found. Veins can be up to 10mm long. Common white opal also occurs as large as a baseball (Downing, 2003).

Figure 7-12: Examples of gems in the Okanagan area. Photo by D. Capstick.

KELOWNA – DIAMOND EXPLORATION CAPITAL

by John D. Greenough

There are no diamond mines in the Okanagan yet but Kelowna is home to one of the most exciting diamond exploration stories in history. In 1977 Charles Fipke set up C.F. Mineral Research Ltd. and began exploring for ore deposits using heavy minerals and indicator minerals. These are minor mineral components contained in soil material, not much larger than a pinhead. They are recovered from samples by employing special laboratory equipment and then studied and analyzed with advanced instruments. Within a few years Fipke had contributed to the discovery of several gold mines in the United States. His innovative exploration techniques identified the first diamond-bearing rocks in British Columbia in 1978.

Secretly, Fipke enlarged his diamond search. Between 1979 and 1989 he and his partners tracked diamond-related indicator minerals across the continent to the Lac de Gras area in the far north. In 1990 a mineral discovery there by Fipke and son Mark convinced Hugo Dummett, the exploration chief for the giant mining company BHP (Broken Hill Properties), to sign a joint venture with Dia Met Minerals Ltd. (Fipke's exploration company). The next year the joint venture discovered the first diamonds in Canada's Northwest Territories and the announcement sparked the "Great Canadian Diamond Rush". It spread across Canada and created the largest "exploration rush" in Canadian history. On October 14th, 1998, BHP Diamonds (renamed BHP Billiton) and Dia Met Minerals Ltd. opened Canada's first diamond mine, the Ekati Mine. Aber Resources and Rio Tinto opened their Diavik diamond mine closeby in early 2003.
As chronicled by several books (e.g. "Fire Into Ice. Charles Fipke & the Great Diamond Hunt" by Vernon Frolick, 1999. Raincoast Books, Vancouver, 354 p.) the diamond discoveries represent a remarkable human triumph and incredible technological success story. Many field and laboratory personnel, including Fipke family members, contributed to the diamond exploration effort. It was based on the inference that during the Ice Age, glaciers moving west from the central Northwest Territories had spread indicator minerals half-way across a continent.

The important indicator minerals related to the discovery were chrome diopside and pyrope garnets, tiny mineral grains that are somewhat jewel-like themselves (Figure 8-12). Some chemical varieties of these mineral grains are more indicative of diamonds than others. John Gurney had figured out a way to chemically "fingerprint" the most important indicator mineral varieties associated specifically with economic diamond "pipes". His information was critical to keeping the search going. In the early 1980's Stu Blusson helped infer the direction of indicator mineral transport. Ultimately, with assistance from a number of other geologists including M. Roed, the indicator minerals led to Point Lake where in 1991 Hugo Dummett and Fipke spotted the drill hole that yielded diamond-bearing kimberlite core.

These discoveries have affected Canada in many ways. The giant Voisey's Bay nickel deposit in Newfoundland-Labrador stemmed from the resultant exploration boom, other prospective diamond pipes were discovered in many parts of Canada, and activity is ongoing.

The Ekati diamond mine produced $2.8 billion dollars worth of diamonds between 1998 and 2002, 3.5 to 4.5 million carats/year. Diavik will produce between 6 and 8 million carats/year! The mines employ approximately 2000 people, many of them native Canadians. The diamond mines, designed to have minimal environmental impact, make a significant contribution to the Canadian Gross Domestic Product and balance of payments. The tax dollars build hospitals, schools and universities.

Three diamond cutting and polishing facilities add $9 million to the Northwest Territories yearly economy.

Within a few years Canada may produce 18% of the world's diamonds making it the third largest producer worldwide. C.F. Mineral Research Ltd. and associated exploration companies continue to provide local employment, and BHP Billiton operates an exploration office staffed with world-class scientists.

Canada's diamond legacy started in Kelowna, created wealth for local Dia Met shareholders, and will impact this community, and Canada, far into the future.

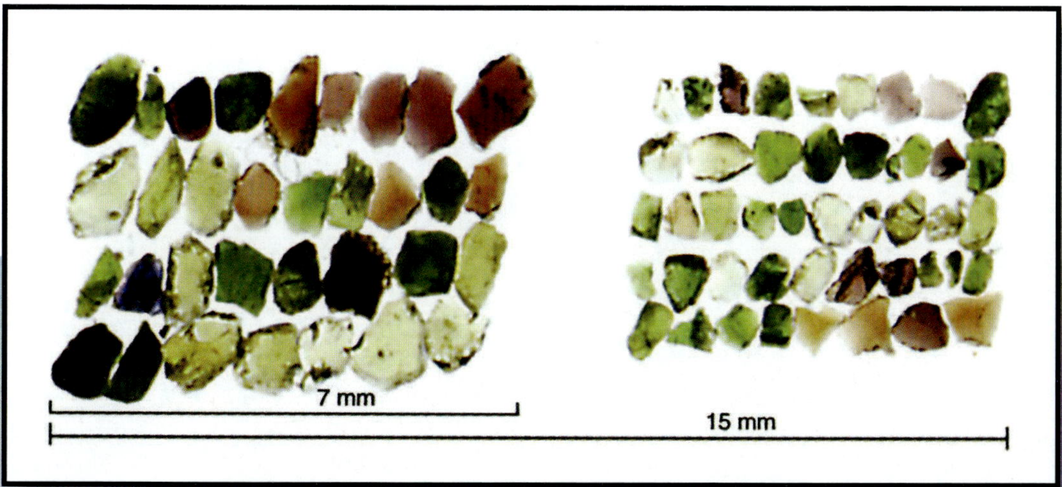

Figure 8-12: Diamond indicator mineral grains mounted in a plastic plug, polished, and ready for chemical analysis. Their compositions will determine whether geologists are on the trail to another diamond mine. Red grains are garnets, dark greens are diopside and the faint greenish yellow grains are olivine. Width of photograph = 15 mm. Sample courtesy of C.F. Mineral Research.

Figure 9-12: Layer Cake Hill and the "Pinnacle" in Gallaghers Canyon, Mission Creek Valley
Acrylic painting by M. Roed.

CHAPTER 13

HYDROLOGY AND WATERSHED MANAGEMENT

by Don Dobson

Hydrologically Kelowna is situated in the Okanagan Lake watershed (Figure 1-13). A watershed is defined as the land area that accumulates all the runoff (surface and groundwater) above a specified point. The southern boundary of the Okanagan Lake watershed is the outlet of the lake at Penticton, and the northern boundary is just north of Armstrong. The area of the entire Okanagan Basin watershed is 6,187 square kilometres, one third the size of Switzerland. The City of Kelowna encompasses the lower portions of several streams that flow into Okanagan Lake. Mission Creek is the largest tributary in the Okanagan Lake watershed, and has an area of 900 square kilometres. It accounts for 15% of the entire Okanagan Basin watershed. The other main streams in the Kelowna area are: McDougall Creek, Powers Creek, Trepanier Creek, Lambly (Bear) Creek, Kelowna Creek, Bellevue Creek, Brandt Creek and Lebanon Creek. The main streams at Penticton include Penticton, Ellis and Shingle Creeks – at Vernon, Coldstream, Vernon and Fortune Creeks (Figure 1-13).

WATER SOURCE

The Okanagan Valley is part of the Interior Dry Belt. Annual precipitation over the Okanagan Basin averages approximately 647 mm, the majority of which falls as snow. Although precipitation for the period June - October averages 237 mm, most of it evaporates before it can runoff. Unless water from the melting snow is stored for use during the dry summer months, there would be insufficient water in the streams to meet domestic and irrigation demands. When snow is the primary source of water to a stream, approximately 90% of the annual runoff volume occurs during the period May-July when the snow is melting (Figure 2-13).

WATER CYCLE

By reviewing the normal consumption of the water for the Okanagan Basin as illustrated in Figure 3-13, it becomes evident how little water there really is available. Of the 647 mm of precipitation that falls in the basin, 524 mm is used by the forests and other vegetation for evapotranspiration, 45 mm infiltration to groundwater, and 78 mm flows into Okanagan

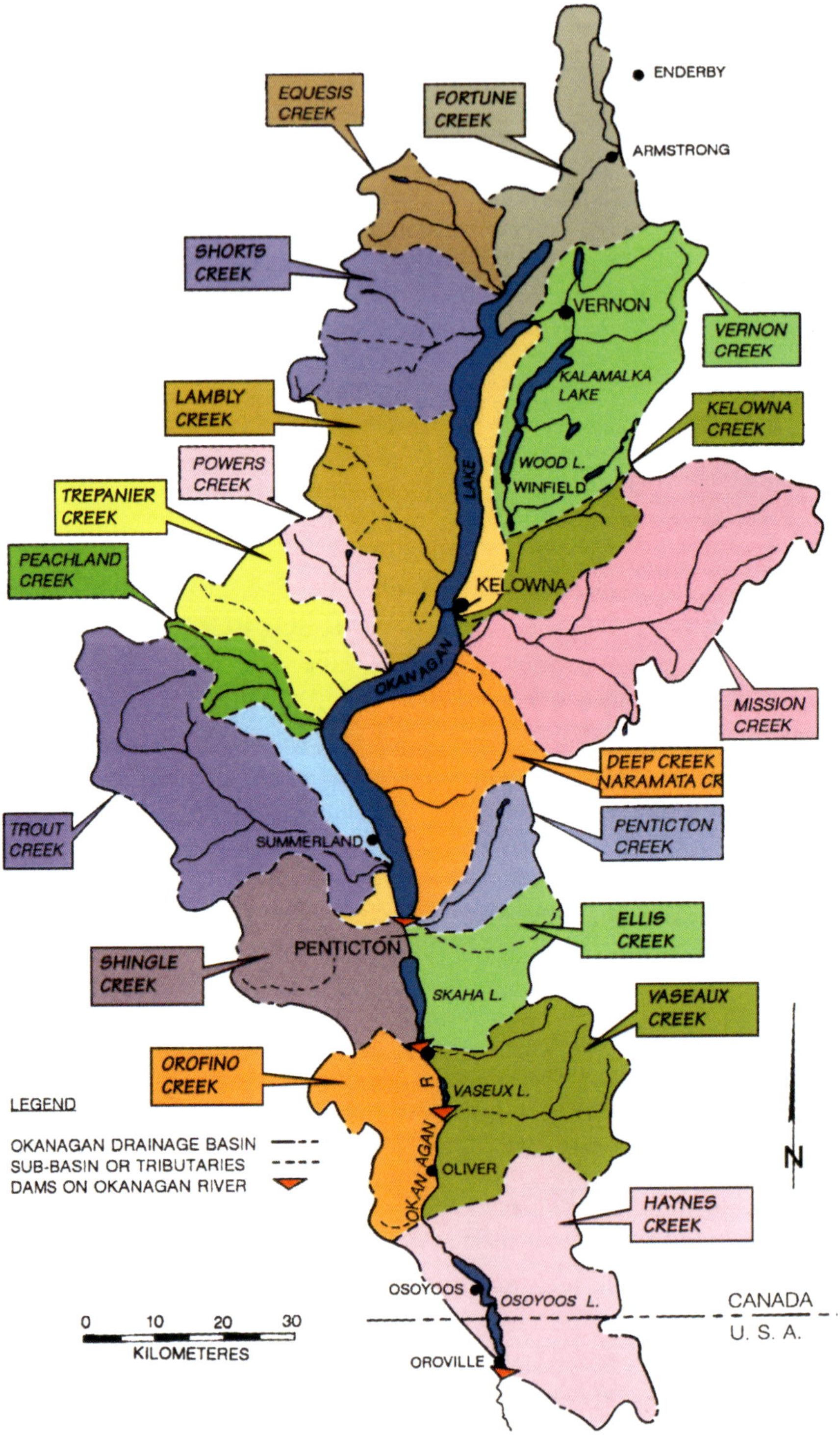

Figure 1-13: The Okanagan Basin Watershed has an area of 6,187 square kilometres and receives 647 centimetres of precipitation a year. The watershed is divided into seventeen sub-basins, the largest of which is Mission Creek than accounts for 15% of runoff.

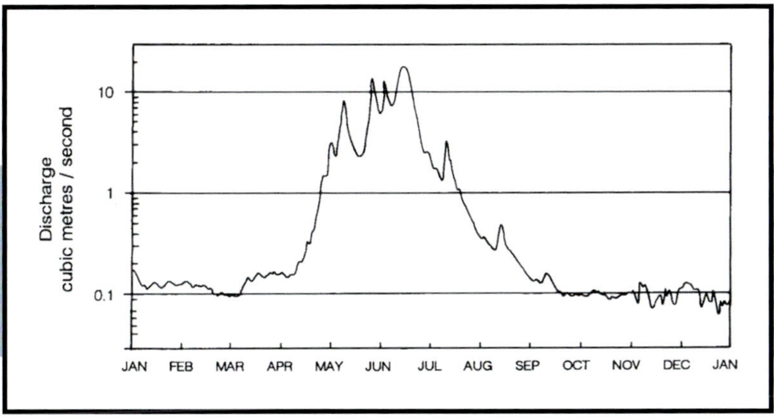

Figure 2-13: A typical hydrograph of a stream such as Mission Creek in the Interior of British Columbia.

Lake of which 25 mm evaporates. This leaves only 20 mm (3%) of the surface water to be used for domestic or irrigation purposes (Figure 3-13).

WATER USE

Water required to meet the daily needs of the City of Kelowna is obtained either from Okanagan Lake or from several of the local creeks. Black Mountain Irrigation District has water licenses in the Mission Creek watershed with reservoirs (Ideal Reservoir, Fish Hawk Reservoir, Graystoke Reservoir) located in the uplands that store 15.4 million cubic metres. Southeast Kelowna Irrigation District has water licenses in the Hydraulic Creek watershed with reservoirs (Hydraulic Reservoir, Browne Reservoir, Turtle Reservoir) that store 16.3 million cubic metres. Glenmore-Ellison Irrigation District has the rights to store 6.9 million cubic metres of water in reservoirs (Postill Reservoir, Bulman Reservoir, South Reservoir) in the Kelowna Creek watershed. Another major supplier is Rutland Waterworks that uses groundwater as the source of water for its users in the Rutland area (refer to Chapter 13 for additional information on groundwater).

MANAGEMENT

Surface water is now managed by the ministry of Water, Land and Air Protection whereas water licensing is provided by Land and Water BC Inc.. Water may be used from a stream or lake only after a water license is issued from the government. The reservoirs in most cases were pre-existing lakes where the volume of water that could be stored was increased by the construction of a dam at the outlet, with control works to regulate the amount of water released. Typically, water is stored in the reservoir as the snow melts during May and June, until the reservoir is full. Water is then released according to the needs for domestic and irrigation uses over the July - October period. Depending upon the system, irrigation can utilize up to 80% of the water supplied by a water district. In a dry year nearly all the water in storage will be used by the end of October.

MULTIPLE USE WATERSHEDS

Land in the watersheds is managed for multiple uses. Most of the uplands of the watersheds are Crown land, meaning that the land is owned by the province. The lowlands or valley bottomlands are all private

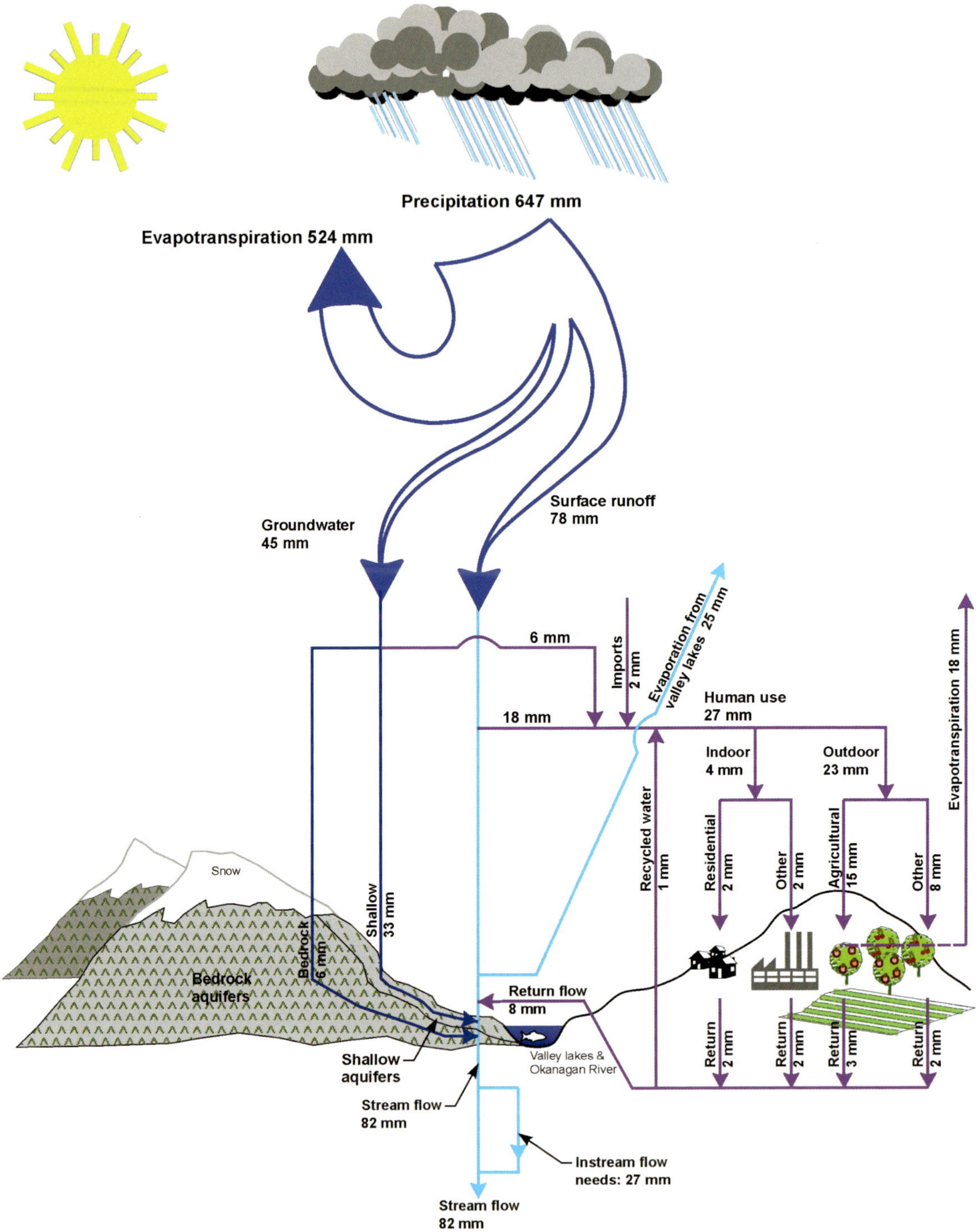

Figure 3-13: The Water Cycle in the Okanagan Basin. Courtesy of the Okanagan Water Board.

land. The multiple use philosophy provides for a wide variety of uses on private land, ranging from agriculture to industrial and residential developments. On Crown land the licensed users include cattle grazing, forest development, mineral development, water resource development, guiding and trapping. Another important but not licensed use is recreation.

EFFECT OF MODERN LAND USE

Land use practices can have significant impacts on the quality and quantity of the water that runs off the land. Timber harvesting is the dominant activity on Crown land, where the majority of the runoff is derived. Forest development is carried out by timber companies such as Tolko Industries Ltd., Weyerhaeuser Canada Ltd., Gorman Bros. Lumber Ltd. and BC Timber Sales in accordance with policies of the Ministry of Forests. Tree species range from Interior Douglas fir and Ponderosa pine which occur on the lower slopes above the valley floor, to Lodgepole pine, Engelmann spruce, and Western larch which dominate the upland. Most harvesting over the past two decades has occurred on the upland and focused on the Lodgepole pine stands which have been infested with mountain pine beetle.

WATER QUANTITY

Logging can affect the hydrology of the watershed by changing the quantity and timing of runoff. As the forest cover is removed more snow accumulates on the ground. Up to 30% more water can occur in the form of snow in a clear-cut opening compared to that in an undisturbed forest. This increase in water becomes significant when the amount of forest cover removed exceeds 20% of the watershed area. It has also been determined, through research, that snow in the clear-cut openings melts approximately two weeks earlier than in the natural forest.

WATER QUALITY

Water quality can also be affected by forest development activities. Soil disturbance from roads, landings and skid trails can increase the amount of sediment that reaches a stream. However, this impact can be controlled through careful construction and maintenance procedures. The water quality may also be affected by increased runoff volumes in the streams resulting from removal of the forest cover and erosion in the stream channel. This impact can be controlled also by planning the rate at which the forest cover is removed so as to maintain close to natural flows in the stream. Waterborne diseases such as

Acrylic painting of Traders Cove, Okanagan Lake, by M.Roed.

cryptosporidiosis and giardiasis have become endemic in most of the valley watersheds. The cysts of these gastro-intestinal illnesses are carried by warm blooded animals such as deer, beaver, cattle and humans. Water purveyors continue to upgrade their water treatment facilities in order to provide safe drinking water. The provincial government passed the Drinking Water Protection Act in 2001 that is intended to improve the protection of domestic water supplies. Work to implement this new legislation is underway.

VITAL RESOURCE

All of the watersheds within the Okanagan basin, including those in the Kelowna area, are vital to the environmental sustainability of the basin. Through increased awareness of the impacts of human activities in the watersheds and knowing that for every action there is a reaction, the environmental quality can be protected. Whether it is the intense development on the valley floor or activities on Crown land, the priority should always be to protect the water resource.

SUMMARY

It is important to be aware that only 12% of the total water that falls on the basin appears as runoff, and only 3% is available for human uses. Since water has always been a scarce resource in the Okanagan Valley, it should be managed with care. With populations continuing to increase, the demand for clean water will also increase. Based on the fact that the amount of available water is limited, optimal management of the resource will be essential. To accomplish this will require additional planning on a watershed basis. Detailed watershed management plans will be required that incorporate forest management on Crown land and urban water management on private land. Management in isolation, that is, Crown land managed separately from private land, and vice versa cannot continue. Water is the life blood of this valley.

Figure 4-13 (front cover): Hardy Falls, Peachland, B.C., acrylic painting by M. Roed. Hardy Falls is situated several hundred metres off Highway 97 south of Peachland, British Columbia. The falls are developed on Deep Creek as it struggles to penetrate stubborn, massive layers of granite.

CHAPTER 14

GROUNDWATER RESOURCES

by Murray A. Roed

Groundwater refers to free water that occurs below the surface of the Earth. Its upper surface is referred to as the water table. All lakes, ponds and stream levels more or less define this level. All groundwater originates from precipitation in the form of rain or snow (Figure 3-13). Some water is evaporated, some is used by plants and trees, some runs off into streams, rivers and lakes, and some is retained by soils. Whatever remains, usually less than 7%, sinks into the ground and becomes groundwater. Fractures in rocks and pores between sedimentary sand, gravel, silt and clay particles serve to store groundwater and permit its slow movement. Such subsurface materials or deposits that can store and transmit groundwater are referred to as aquifers, and holes drilled into the aquifer that produce groundwater economically are called water wells.

GROUNDWATER FLOW

The flow of water through an aquifer is controlled by two factors: (1) The ability of the subsurface soil or rock to transmit water, known as hydraulic conductivity, and (2) the slope of the water table which is called the hydraulic gradient. Aquifers collect water in recharge areas which are at high elevation. The amount of recharge possible is related to the infiltration capacity of the surface. For example, more water would seep into a gravel deposit than a clay deposit.

Everyone knows that water flows from high spots to low spots. Thus, groundwater enters in recharge zones at high elevation and flows through one or more aquifers to a lower elevation. Where the land surface is intersected natural springs may occur forming a discharge zone.

ENVIRONMENTAL SENSITIVITY

The location and characteristics of groundwater recharge and discharge zones in watersheds place constraints on land use and development. For example, pollutants allowed to enter a recharge zone, perhaps at a landfill site, may lead to contamination of the entire "downstream" portion of the aquifer. Surface changes in a recharge zone can affect the quantity of water entering the subsurface. For example, removal of forest cover by fire or logging increases the amount of groundwater whereas urbanization and paved surfaces decreases the amount of groundwater. This

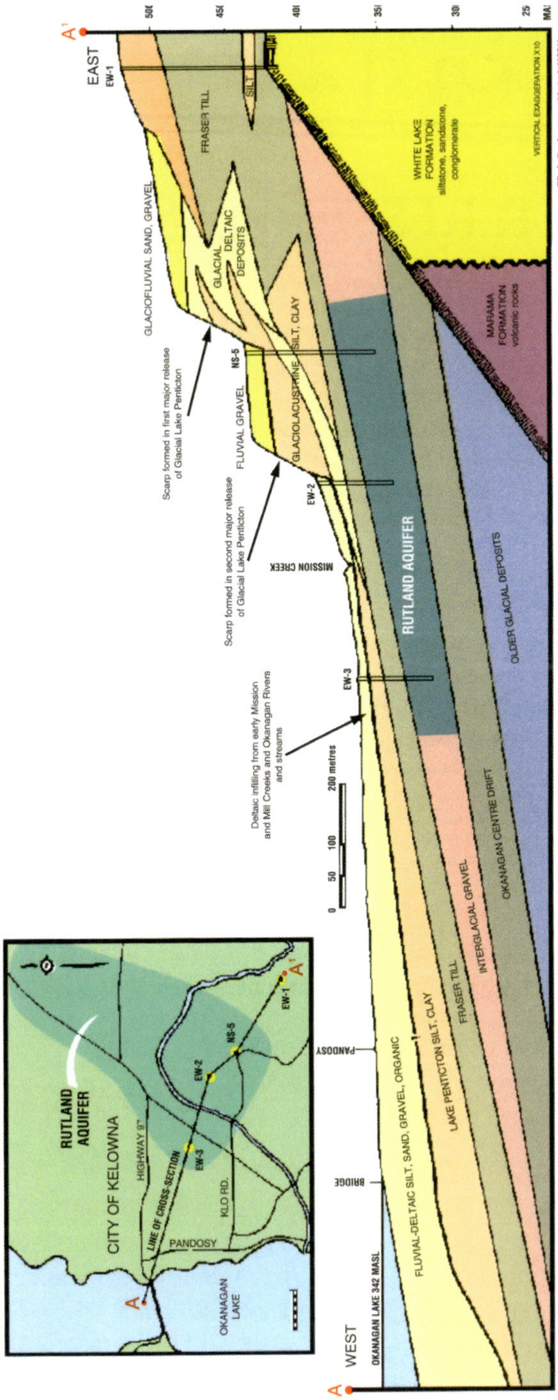

Figure 1-14: Cross-section through the Rutland Aquifer showing the stratigraphy of surficial deposits and location of water wells (EW-3, etc.) employed for the section. An accompanying index map shows the line of cross-section A-A1.

in turn may have an important impact on discharge zones at lower elevations. Landslides often originate in groundwater discharge zones where springs and high pore pressure reduce the strength of soils on a slope.

GROUNDWATER USE IN THE KELOWNA AREA

There are thousands of water wells in the Kelowna area, but unfortunately a comprehensive study of the various aquifers that exist is not available. Historically, groundwater has always been a major component of the city's domestic and agricultural water supply.

MAGNITUDE OF GROUNDWATER USE

The magnitude of groundwater use is illustrated by the Rutland Waterworks District installations. Just below the Rutland fan in the vicinity of Mission Creek, the Rutland Waterworks District has developed 15 wells between 32 and 106 metres in depth that serve 4,500 connections (1991) in the Rutland District. The aquifer is called the Rutland Aquifer (Figure 1-14) and the wells are pumped at rates of 900 to 5,600 litres per minute. Average monthly summer production is 681 million litres, with consumption dropping to 205 million litres in winter (courtesy of Rutland Water District). Total production in 1977 was three billion litres (800,000,000 U.S.gallons), in 1991 it was three and a half billion litres and in 1994 four and a quarter billion litres. Volumes were not appreciably different in 2004.

In comparison, the Black Mountain Irrigation District in 1994 withdrew a total of 125 billion litres from Mission Creek, nearly 30 times the quantity withdrawn from the Rutland aquifer. East Kelowna Irrigation District withdraws similar amounts from a tributary to Mission Creek.

Many water districts in the Okanagan Valley supplement surface water resources with groundwater. All of the rural developments including some subdivisions and some orchards in the area rely on water wells. Numerous golf courses have drilled their own water wells for maintenance of their fairways. Several bottled water factories, at least one brewery, several dairy product factories, and all major cattle ranches and feed lots in the region rely on groundwater resources.

GEOLOGY OF THE RUTLAND AQUIFER

A cross-section through the Rutland aquifer is shown in Figure 1-14. It occurs as an extensive gravel bed 100 metres thick that is overlain by outwash deposits and till inferred to belong to the Fraser glaciation, silt and clay deposits of Glacial Lake Penticton, and fluvial-deltaic deposits which have accumulated since the end of glacial time in the former embayment of Okanagan Lake now occupied by Kelowna.

Data for the construction of this cross-section have been derived from water wells (Lowen and Letvak, 1981), a surficial geology map (Nasmith, 1962), and from 1994 drill hole data at the east end of the floating bridge (supplied through the courtesy of the Ministry of Transportation and Highways).

It is known from regional information that the Rutland aquifer extends northward to at least the airport although the water within it is not of uniform quality. In places it is high in iron and manganese. Remarkably, the water from the aquifer appears to have remained unaffected by agricultural chemicals and residential septic systems in East Kelowna.

How and where this aquifer is recharged is not known at this time. It is possible that a thick gravel exposed in Gallagher's Canyon (Figure 15-9) is part of the aquifer. The base of this outcrop is at an elevation of approximately 420 metres and correlates with a projection of the aquifer's stratigraphic position in the subsurface. If this is the case, then a downstream section of Mission Creek flows directly over the aquifer. It is possible therefore that this portion of the creek may be one of the main points of recharge for this important aquifer.

GROUNDWATER ASSESSMENT PROJECT OF THE OKANAGAN BASIN

The Okanagan's precious groundwater resource is about to receive combined federal and provincial attention. An action program has been outlined by Natural Resources Canada, the Geological Survey of Canada and British Columbia Ministry of Water, Land and Air Protection. This regional groundwater assessment project will be carried out over several years. It is designed to determine the geology of major aquifers, assess their yields, identify principal sensitivities of groundwater recharge areas and review potential impacts on this resource relative to climatic change.

The project is called Groundwater Assessment in the Okanagan Basin (GAOB). Part of the process will be to raise public awareness of the importance of groundwater. It is estimated that there are over 10,000 producing wells in the Okanagan. Future growth will undoubtedly include a great demand for clean potable groundwater as the price of treating and developing surface water rises and as potential reservoir capability in the upland diminishes.

KELOWNA WARM SPRINGS

by John D. Greenough

Few people know about them, but there are warm-springs called the Angel Springs (Figure 2-14) up the KLO Creek valley ([31] in Figure 1-9). They are most accessible by driving up June Springs Road to the 1km sign on the Forestry Service Road and then hiking northeast for about one hour along an old logging road and hiking trail.

Water from the springs crosses the hiking trail in a small ravine, and is easily identified by its bright orange colour. The colour results from exuberant algal growth and the precipitation of iron-bearing minerals, These features plus water temperatures of about 20 to 25 degrees Celsius discourage bathing. However, these temperatures probably reflect mixing of hot-spring water with colder surface and cool-spring waters. Temperatures as high as 29 degrees Celsius occur around seeps a few hundred metres up the ravine.

Perhaps the most fasinating geological aspect of the springs is the extensive tufa deposits that form a series of terraces with 10 metre-high benches at irregular intervals along the ravine. These deposits attest to the presence of prolonged hydrothermal (hot water) activity that could be thousands of years old and that may have been more productive in past times.

At present we can only speculate about the origin of the warm water. Possibly the ravine demarcates a fault in the ancient gneisses that outcrop along the walls of the gorge. Water originating deep in the earth could follow the fault to the surface. Possible heat sources include the earth's interior (temperature rises with depth in the earth) or a magma chamber. The spring does occur within the Kelowna hot spot as shown in Figure 11-10. Another hypothesis is that the water is flowing from gravel deposits underlying plateau basalts that are one to three million years old. The basalts outcrop along the Kettle Valley Railway bed just above the ravine and rumour has it that drill holes through similar basalts in the area encountered warm water in underlying gravel deposits.

Figure 2-14: Acrylic painting by M. Roed showing the terraces of tufa at Angel Springs.

CHAPTER 15

EARTH SCIENCE AND THE FUTURE OF THE OKANAGAN VALLEY

by Murray A. Roed

The growth and application of earth science in the twentieth century has been explosive, but the importance of these developments has yet to be appreciated, especially by local governments in British Columbia. Geology affects every aspect of modern life in the Okanagan. The best agricultural land is developed on a variety of surficial deposits which reflect their geologic origin. Possibly, the Okanagan's unique soils are responsible for the success of grape growing and a thriving wine industry. Mines and mineral development have had a major economic impact and allowed the development of advanced innovative techniques in mining engineering that are now in demand on a world wide scale.

The beauty and variety of the Okanagan terrain is widely known and due to a complex and awesome geologic history. This geologic design work of Mother Nature continues to attract more and more visitors, sustain a diversified tourist industry, and make the Okanagan a preferred place to live and enjoy almost every type of recreational pursuit.

However, entering the twenty-first century, the resources of the Okanagan will have to be known and understood in far more detail than at present. Water management including the preservation and enhancement of existing water quality will be the most challenging task. Far too little is being done about water use issues at this time. This resource is limited and must be managed in optimum harmony with an array of conflicting land use issues, not the least of which is the preservation of natural biological systems. The Kokanee, wildlife, forest, and many of our basic needs all depend on wise and appropriate water management.

WATERSHED MANAGEMENT

Serious watershed management is now a government mandate in British Columbia. The long term prospect for greatly improved watershed management, better water quality, and healthier forests can be envisaged. However, management of both Crown and private land has to be integrated.

Requirements for earth science, hydrology and engineering expertise in the forest industry and water management has driven demand for expertise. This demand will increase and extend into the future which bodes well for young scientists of the province.

CONSTRUCTION MATERIALS

It has been shown that modern geotechnical engineering can be applied to build larger and higher structures in the Okanagan. The highway system is gradually being improved, large rural and suburban developments are under construction, major industries are emerging, and many construction materials are required. All of this activity involves concrete and pavement, products of the aggregate industry, and in turn the existence of accessible aggregate resources. At the rate sand and gravel deposits of both fluvial and glacial origin are being used for building sites, the Okanagan will soon be depleted of this valuable resource, yet there appears to be little awareness or concern for this.

RESIDENTIAL DEVELOPMENT

Residential development in the future is projected to depend on major capital expenditures for sewage treatment. Although controversial, development using individual septic systems in thick suitable surficial deposits where terrain is well drained can be an economical and environmentally safe option. Through modern geologic and hydrogeologic studies, it is feasible to locate terrain that is ideally suited for individual sewage systems without jeopardizing the environment. Failure of septic systems is rooted in poor evaluation and siting techniques and construction methods of the past. There are more data supporting the use of tiles fields with attendant environmental integrity than not, a good example of this is the Rutland aquifer and how it has remained uncontaminated throughout the years.

URBAN AND REGIONAL PLANNING

Urban and regional planners are beginning to respond to some concerns regarding storm water management, hazard land evaluation, and flooding, but a suitable and detailed terrain and hydrologic data base for evaluation of these is lacking in the Okanagan. Compared to other places in the world, even in Canada, our almost nonexistent database for these important terrain characteristics is insufficient. There has to be a long term commitment to improve.

Figure 1-15: The vastness and beauty of the Okanagan landscape is captured in this aerial photo looking north. Peachland is to the left, the fire ravaged terrain of Okanagan Mountain Park is on the right and Okanagan Lake remains blue and beautiful. Let's keep it that way! Photo by John D. Greenough, 2004.

MINERAL RESOURCES

Minerals such as gold, silver, lead, zinc, copper and molybdenum still lie undetected in the bedrock of the Okanagan landscape. According to recent work and new theories involving plate tectonics, the diversified bedrock terrain west of Okanagan Lake offers the best prospects for new discoveries. Volcanic deposits of the Kelowna caldera complex also offer possibilities for rare minerals such as opal and agate, but little exploration has been conducted. The impetus for encouraging mineral exploration lies with government, and a good start would be comprehensive and detailed bedrock and surficial geology maps of the Okanagan Valley.

SUMMARY

If a conclusion to this book can be written, it has to be that there are substantial opportunities, unprecedented in history, for students of earth science in the growth and well being of the Okanagan. Politicians who are aware of our critical resource needs can play a leading role in channeling sufficient funds to enhance the resource data base. The academic community should be encouraged to expand teaching and research in all phases of earth science and engineering. A central well stocked science library is critically required, and outlets for remote sensing products, maps and government science publications are necessities. Geotechnical engineering professionals should begin to inventory soils, store excavating and drilling results and promote related research to help build a resource data base for public use. Each Regional District in the Okanagan should initiate this activity. Residents have a role to play in learning to appreciate the usefulness and/or sensitivity of the resources and be reasonably tolerant in land use conflicts. Planners have to learn more about how to economically utilize the fruits of earth science in a socially agreeable, environmentally acceptable and economy-sustaining manner. There is much to be done. Above all, get outside and enjoy this beautiful part of British Columbia!

Figure 2-15: Squally Point, Okanagan Lake, acrylic painting by M.Roed The most dramatic bedrock cliffs along Okanagan Lake are located near Squally Point in this westerly view towards Peachland. The rocks belong to the Precambrian/Devonian aged Monashee Metamorphic complex and are granitic gneisses intruded in places by large plutons of granite that are Tertiary and Mesozoic in age.

Figure 3-15: Early morning sun on Silt Bluffs, Glacial Lake Penticton near Sun Oka Beach Provincial Park. Acrylic painting by M. Roed.

REFERENCES

Agriculture Canada, 1993. Soil Landscapes of Canada, British Columbia-South. Canada Soil Inventory, Centre for Land and Biological Resources Research, Agriculture Canada, Contribution Number 89-03.

Alley, N.F., 1976. The Palynology and Paleoclimatic Significance of a Dated Core of Holocene Peat, Okanagan Valley, Southern British Columbia, Canadian Journal of Earth Sciences, v. 13, 1131-1144.

Armstrong, R.L., Parish, R.R., van der Heyden, P., Scott, K., Runkle, D., and Brown R.L., 1991. Early Proterozoic basement exposure in the southern Cordillera; core gneiss of Frenchmand Cap, Unit 1 of the Grand Forks Gneiss, and the Vaseux Formation. Canadian Journal of Earth Sciences, v. 28, 1169-1201.

Bailey, J. and Rousseau, M.K., 1994. An Archeological assessment of prehistoric Site DkQw 36 in Okanagan Mountain Park, near Naramata, south-central B.C. Report on file at the Archeologic Branch, Victoria, B.C.

Baker, J. (Editor), 1975. Okanagan College Archeological Research Project, Okanagan College, Kelowna, B.C.

Baker, J., 1990. Archeological research concerning the origins of the Okanagan people. In Okanagan Sources, Editor Jean Webber, Penticton, B.C., Theytus Books Ltd.

Borden, C.E., 1979. Peopling and Early Cultures of the Pacific Northwest: A view from British Columbia, Canada. Science v. 203, 1963-1971.

Buchannan, R.G., and Haughton, D.R., 10978. Physical, chemical, mineralogical properties of B.C. glaciolacustrine silts and fundamental design implications. Ministry of Transportation and Highways, Vitoria, B.C.

Cairnes, C.E., 1932. Mineral resources of northern Okanagan Valley, British Columbia. Geol. Surv. Canada, Summary Report 1931, Part A, 66-109.

Cairnes, C.E., 1937. Mineral deposits of the West Half of Kettle River area, British Columbia. Geol. Surv. Canada, Paper 37-21.

Carr, J.M., 1967. Geology of the Brenda Lake area. B.C Minister of Mines and Petroleum Resources, Ann. Rept., 183-212.

Cass, D.E., Kenning, B.F.I, and Rawlings, G., 1992. The Philpott Road Debris Failure, B.C.1990: The Impacts of Geology, Hydrology and Logging Activities. Geotechnique and Natural Hazards; a Symposium sponsored by the Vancouver Geotechnical Society and the Canadian Geotechnical Society, May 6-9, 1992, 319-325.

Chatwin, S.C., Howes, D.E., Schwab, J.W., and Swanson, D.N., 1994. A Guide for Management of Landslide-prone Terrain in the Pacific Northwest, Second Edition. Research Program, Ministry of Forests, British Columbia, 220p.

Chritstopher, P.A., 1977. Uranium mineralization in the Hydraulic Lake area, British Columbia; in Geological Fieldwork 1976, B.C. Department of Mines and Petroleum Resources, 11-14. Church, B.N., 1980. Geology of the Tertiary Outlier (West Half, NTS 82E). B.C Ministry of Mines and Petroleum Resources, Preliminary Map 39.

Church, B.N., 1981. Geology of the Kelowna Tertiary Outlier (East Half, NTS 82E). B.C Ministry of Mines and Petroleum Resources, Preliminary Map 45.

Clowes, R.M., Zelt, C.A., Amor, J.A., and Ellis, R.M., 1995. Lithospheric structure in the southern Canadian Cordillera from a network of seismic refraction lines, Canadian Journal of Earth Sciences, v. 32, 1485-1513.

Cole, D., and Lockner, B., 1989. The Journals of George M. Dawson, British Columbia, 1875-1878. Vol. I, 1875-1876, 296p; v .II, 1877-1878, 297-611p. University of British Columbia Press. Copp, S., 1980. A dated pictograph from the South Okanagan Valley, British Columbia. Canadian Rock Art Research Associates Newsletter, 14: 44-48.

Corner, J., 1968. Pictographs: Indian Rock paintings. Vernon, B.C.

Daughtry, K.L., Thompson, R.I., and Erdmer, P., 2000. New field studies of the Chapperon Group, Vernon southwest maps area, British Columbia. Geol. Survey of Canada, Current Research 2000-A19, 4p.

Dawson, G.M., 1878. Explorations in British Columbia. Geol. Survey of Canada, Report of Progress 1876-77, 16-149.

Dawson, G.M., 1879. Report on exploration in the southern portion of British Columbia. Geol. Survey of Canada, Report of Progress 1877-1878, Pt.B, 157-158.

Dawson, G.M., 1879. Preliminary report on the physical and geological features of the Interior of British Columbia. Geol. Survey of Canada, Report of progress, 1877-1878.

Davis, R., Chilton. R., Ottenbreit, L., Scheeler, M., Vielvoye, J., Williams, R., and Wittneben, U., 1984. Atlas of suitable grape growing locations in the Okanagan and Similkameen valleys of British Columbia. Agriculture Canada, Summerland, B.C., 141p.

Dayton and Knight Ltd., 1989. City of Kelowna, Stage 1, Master Drainage Plan. Consultant's Report.

Downing, P.B., 1993. Okanagan Opal. Lapidary Journal, v. 16, No. 11, 63-66.

Downing, Paul B., 2003. Beautiful British Columbia, Rock and Gem, October, 2003, 20-30.

Eggers, N. Greenough, J.D., and Cernak, T., 2004. Classification of Okanagan Chardonnay Wines by origin using volatile components. Geoscience Canada.

Evans, J., 1993. The Okanagan Lake Bridge. Kelowna Centennial Museum Association, Kelowna, B.C. V1Y 6S7, 11p.

Evans, S.G., 1982. Landslides and surficial deposits in urban areas of British Columbia: a review. Canadian Geotechnical Journal, v. 19, No. 3, 269-288.

Eyles, N., Mullins, H.T., and Hine, A.C., 1990. Thick and fast: Sedimentation in a Pleistocene fiord lake of B.C., Canada. Geology, v. 18, 1153-1157.

Fairbank, B.D., and Faulkner, R.L., 1992. Geothermal resources of British Columbia. Geol. Survey of Canada, Open File 2526, map, scale 1:2,000,000.

Falk, L., 1982. Hiking trails in the Okanagan. Mosaic Enterprises, 1460 St. Paul Street, Kelowna, B.C., 60p.

Fladmark, K.R., 1983. Times and places: Environmental correlates of Mid-to-Late Wisconsin human population expansionin North America. InEarly man in the New World. Sage Publications, Beverly Hill, California.

Fladmark, K.R., Driver, J.C., and Alexander, D., 1988. The Paleoindian component at Charlie Lake Cave (HbRf 39), B.C. American Antiquity, v. 53, 371-114.

Flint, R.F., 1935. "White Silt"deposits in the Okanagan Valley, B.C. Roy. Society Canada Transactions, v. 29, 107-114.

Flint, R.F., 1935. Glacial features of the southern Okanagan. Geol. Society America Bull., v. 46, 169-193.

Frolick, Vernon, 1999. Fire into ice. Charles Fipke & the Great Diamond Hunt. Raincoast Books, Vancouver, 354p.

Fulton, R.J., 1965. Surficial Geology-Vernon. Geol. Survey of Canada, Map, Scale 1:126.720.

Fulton, R.J., 1965. Silt Deposition in Late-Glacial Lakes in Southern British Columbia. Am. Jour. Sci., v. 263, 553-570.

Fulton, R.J., 1969. Glacial Lake History, Southern Interior Plateau, British Columbia. Geol. Survey. Canada, Paper 69-37, 14p.

Fulton, R.J., 1975. Quaternary geology and geomorphology, Nicola-Vernon, British Columbia (82L W1/2 and 921 E1/20) Geol. Survey of Canada, Memoir 380, 50p.

Fulton, R.J. and G.W.Smith, 1978. Late Pleistocene Stratigraphy of South-Central British Columbia. Canadian. Journal of Earth Sciences, v.15, 971-980.

Fulton, R.J., Irving, E., and Wheadon, P.M., 1992. Stratigraphy and Paleomagnetism of Brunhes and Matuyama (>790 ka) Quaternary Deposits at Merritt, British Columbia. Canadian. Journal of Earth Sciences., v. 29, no.1, 76-92.

Fulton, R.J., 2003. Geology of Okanagan Valley wine region of British Columbia. (Abstract) 2003 Seattle Annual Meeting of the Geological Society of America.

Gabrielse, H., Monger, J.W.H., Wheeler, J.O. and Yorath, C.J., 1991. Tectonic framework; Part A, Morphological belts, tectonic assemblages and terranes. In Geology of the Cordilleran Orogen in Canada. Edited by H. Gabrielse and J.W.H. Monger. Geological Survey of Canada, Geology of Canada, v. 4, 15-28.

Godwin, C.I., Watson, P.H., and Shen, K., 1986. Genesis of the Lass Vein System, Beaverdell Silver Camp, South-Central British Columbia. Canadian Journal of Earth Sciences, v. 23, 1615-1626.

Greenough, J.D., Longerich, H.P., and Jackson, S.E., 1975. Element fingerprinting of Okanagan Valley wines using ICP-MS: Relationships between wine composition, vineyard and wine colour. Australian Journal of Grape Wine Research, v. 3, 75-83.

Greenough, J.D., Longerich, H.P., and Jackson, S.E., 1996. Trace element concentrations in wines by ICP-MS: Evidence for the role of solubility in determining uptake by plants. Canadian Journal of Applied Spectroscopy, v. 41, 76-80.

Greenough, J.D., and Owen, J.V. 1998. Ingneous Layering in a Dacite: On the Origin and Significance of Layer Cake Mountain, Kelowna, B.C., Canada. Mineralogical Magazine, v. 62, 731-742.

Harris, G.R., Hatfield, H., and Tassie, P., 1989. The Okanagan Brigade Trail in the south Okanagan. Wayside Press, Vernon, B.C.

Harris, G.R., 1992. Mine Closure – Brenda Mines Ltd. (paper delivered to the Mining Association of Canada.)

Henrichs, M.L., Hebda, R.J., and Walker, I.R., 2001a. Halocene vegetation and natural disturbance in the Englemann Spruce – Subalpine Fir biogeoclimatic zone at Mt. Kobau, B.C. Canadian Journal of Forest Research v. 31: 2183-2199.

Heinrichs, M.L., Walker, I.R., and Mathewes, R.W., 2001b. Chironomid-based paleosalinity records in southern British Columbia, Canada: a comparison of transfer functions. Journal of Paleolimnology v. 26:147-159.

Hickson, C.J., 1992. Volcanism in the Canadian Cordillera: Should We Worry? Geotechnique and Natural Hazards. BiTech Publishers Ltd., Vancouver, B.C., 31-40

Holland, S.S., 1964. Landforms of British Columbia, a Physiographic Outline. B.C. Dept. of Mines and Pet. Res., Bull. 48, 138p.

Howie, J.A., and Jinks, A.R., 1994. The Grand Okanagan Lake Front Resort and Conference Centre, Kelowna, B.C. 47th Canadian Geotechnical Conference, 1994, Halifax, N.S., 200-209.

Hudson, D.R., 1990. The Okanagan Indians of British Columbia. In Okanagan Sources. Ed. Jean Webber. Penticton, B.C. Theytus Books Ltd.

Jones, A.G., 1959. Vernon map area British Columbia. Geological Survey of Canada, Memoir 296.

Keyser, J.D., 1992. Indian Rock Art of the Columbia Plateau. University of Washington Press. Seattle.

Little, H.W., 1961. Kettle River, West Half, B.C. Geol. Surv. Canada. Map 15-1961.

Lowen, D.A., and Letvak, D.B., 1981. Report on Groundwater-Surface Water Interrelationship, Lower Mission Creek, B.C. B.C Min. Envir., Water Management Branch, Victoria, B.C. 19p.

Lundy, A. (dePfyffer), and Zoellner, D. (Whitham), 1990. Tours Made Easy. R.R.#4, 4881 Lakeshore Road, Kelowna, B.C. V1Y 7R3, 102p.

Macauley, H.A., Hobson, G.D., and Fulton, R.J., 1972. Bedrock Topography of the North Okanagan Valley and Stratigraphy of the Unconsolidated Fill. Geological Survey of Canada, Paper 72-8, 17p.

Malcolm, G., 1995. New Life for Brenda Mines. Mining Magazine, London, England. 271-219.

Mathews, W.H., 1944. Glacial Lakes and Ice Retreat in South Central British Columbia. Roy. Soc. Canada Trans., v. 38, sec. 4, 39-57.

Mathews, W.H., 1988. Neogene Geology of the Okanagan Highland, British Columbia. Canadian Journal of Earth Sciences, v. 25, no. 5, 725-731.

Meyers, R.E., and Taylor, W.A., 1989. Lode Gold-Silver Occurrences of the Okanagan Region, South-Central British Columbia (82E/W, 82L/SW). B.C. Min. Energy, Mines and Petrol. Res., Geol. Surv. Br., O.F. 1989-5.

Ministry of Forests, 1990 Investigation into the causes of the Destructive Debris Flow, Joe Rich-Belgo Creek Area, June 12, 1990. B.C. Min. Forests, Forest Service Investigation Team, Penticton Forest District, 57p.

Nasmith, H., 1962. Late Geological History and Surficial Deposits of the Okanagan Valley, B.C. B.C. Min. Energy, Mines and Petrol. Res., Bull 46, 46p.

Nyland, D., and Miller, G.E., 1977. Geologic Hazards and Urban Development of Silt Deposits in the Penticton Area. B.C. Ministry of Highways and Public Works, Geotechnical Materials Branch.

Okulitch, A.V., 1979. Geology and Mineral Occurrences of the Thompson-Shuswap-Okanagan Region, South-Central British Columbia. Geological Survey of Canada, O.F. 637.

Paynter, S., 1991. First Time Around. Published by the Author, Box 166, West Kelowna, B.C. V0H 2A0, 136p

Pegusch, W., 1957. The Kelowna Floating Bridge. Eng. Jour. April, 1957. 413-421.

Peto, P., 1973. Petrochemical Study of the Similkameen Batholith, British Columbia: Geol. Soc. Amer. Bull. v. 84, 3977-84.

Read, P.B. and Okulitch, A.V., 1977. The Triassic unconformity of South-Central British Columbia. Canadian Journal of Earth Sciences, v. 14, 606-638.

Reinecke, L., 1915. Physiography of the Beaverdell Map Area of the Southern Part of the Interior Plateau of British Columbia. Geol. Surv. Canada, Mus., Bull. no. 11.

Rice, H.M.A., 1960. Geology and Mineral Deposits of the Princeton Map Area, B.C. Geol. Survey., Mem. 243, 136p.

Richards, T.H. and Rousseau, M.K., 1987. Late Prehistoric Cultural Horizons on the Canadian Plateau. Simon Fraser University, Department of Archaeology Publication 16, Burnaby, B.C.

Roed, M.A., 1967. Athabasca Valley Erratics Train and Ice Movement across the Continental Divide. Canadian Journal of Earth Sciences, v. 4, 625-632.

Roed, Murray A., 1975. Cordilleran and Laurentide Multiple Glaciations, West-Central Alberta, Canada. Canadian Journal of Earth Sciences, v. 12, 1493-1515.

Roed, M.A., 1991. Scenic Canyon Regional Park: Natural History Unfolds Here. Okanagan Life, Summer, 1991, p. 11 and 18.

Roed, M.A., 1995. Geology of the Kelowna Area and Origin of the Okanagan Valley, British Columbia Kelowna Geology Committee, 183p.

Roed, M.A. and Greenough, John D., editors, 2004. Okanagan Geology, British Columbia, Kelowna Geology Committee, 220p.

Roed, Murray A. and Fulton, Robert J. editors, 2011. Okanagan Geology South. Okanagan Geology Committee, 238p.

Rosenberg, S.M., Walker, I.R., Mathewes, W.H., and Hallett, D.J., 2004. Midge-inferred Holocene climate history of two subalpine lakes in southern British Columbia, Canada. The Holocene v. 14:258-271.

Ryder, J.M. and Thompson, B., 1986. Neoglaciation in the Southern Coast Mountains of British Columbia: Chronology Prior to the Late Neoglacial Maximum. Canadian Journal of Earth Science, v. 23, no. 3, 273-287.

Sanford, B., 1978. McColloch's Wonder, The Story of the Kettle Valley Railway. Whitecap Books, Vancouver, B.C. 260p.

Soregaroli, A.E. and Whitford, D.F., 1974. Brenda. CIM Special Volume No. 15, Part B, 186-94.

Shaw, J. and Archer, J., 1979. Deglaciation and Glaciolacustrine Conditions, Okanagan Valley, British Columbia, Canada in Moraines and Varves (INQUA Symposium), Rotterdam, Bakema, A.A., 347-355.

Shewchuk, M., 1992. Okanagan Country, An Outdoor Recreation Guide. Sonotek Publishing Ltd., P.O. Box 1752, Merritt, B.C. V0K 2B0, 176p.

Smith, K., 1990. The West Wall Failure at Brenda Mines. Brenda Mines Ltd.

Taylor, V.F., Longerich, H.P., and Greenough, J.D., 2002. Geology and Wine 5. Provenance of Okanagan Valley Wines, British Columbia, using trace elements: promise and limitations. Geoscience of Canada, v. 29, 110-120.

Taylor, V.F., Longerich, H.P., and Greenough, J.D., 2003. Multielement analysis of Canadian vines by inductively coupled plasma mass spectrometry (ICP-MS) and multivariate statistics. Journal of Agricultural and Food Chemistry, v. 51, 856-860.

Templeman-Kluit, D.J. and Harakal, J.E., 1986. Extension across the Eocene Okanagan Crustal Shear in Southern British Columbia. Geology, v. 14, 318-321.

Templeman-Kluit, D.J., 1989. Geological Map with Mineral Occurrences, Fossil Localities, Radiometric Ages and Gravity Field for Penticton Map Area (NTS82E), Southern British Columbia. Geological Survey of Canada, O.F. 1969.

Turner, N., Bouchard, R., and Kennedy, D., 1980. Ethnobotany of the Okanagan-Colville Indians of British Columbia and Washington. British Columbia Provincial Museum Occasional Papers Series, No 21.

Unterschultz, J.L.E., Creaser, R.A., Erdmer, P., Thompson, R.I., and Daughtry, K.L., 2002. North American margin origin of Quesnel Terrane strata in the Southern Canadian Cordillera; inferences from geochemical and Nd isotopic characteristics of Triassic metasedimenatary rocks. Geological Society of American Bulletin, v. 114, 462-475.

Walker, I.R., Reavie, E.D., Palmer, S., and Nordin, R.N., 1993. A palaoenvironmental assessment of human impact on Wood Lake, Okanagan Valley, British Columbia. Quaternary International 20: 51-70.

Walker, I.R. and Pellat, M.G., 2003. Climate change in coastal British Columbia – a paleoenvironmental perspective. Canadian Water Resources Journal 28: 531-566.

Watson, E. and Luckman, B.H., in press. Tree-ring based reconstructions of precipitation for the Southern Canadian Cordillera. Climatic Change.

Wilson, R.J.S. and Carlson, C., 1980. The Archaeology of Kamloops. Simon Fraser University, Department of Archaeology Publication 7, Burnaby, B.C.

Wilson, R.J.S. and Luckman, B.H., 2003. Dendroclimatic reconstruction of maximum summer temperatures from upper treeline sites in interior British Columbia, Canada. The Holocene 13: 851-861.

Wittneben, U., 1986. Soils of the Okanagan and Similkameen Valleys, B.C.

Crawford Falls is located a short hike along a trail above Bellevue Creek near the Crawford Estates subdivision in East Kelowna. It is in a dangerous and nearly inaccessible part of the valley. Obtaining a vantage point to look at the falls is therefore a risky activity and is not recommended. There are two sections to the falls, an upper falls, which is painted here, and a lower less impressive falls lower down the valley (see page 203). The upper falls plunge at least 20 metres straight down into a steeply eroded rock-bounded bowl. The water spills over a ledge of Monashee granitic gneiss. During low water it is possible but risky to stand on this ledge and look down the falls. The falls owe their origin to details of the deglaciation in the area. Possibly 12,000 years ago Bellevue Creek was a raging torrent of meltwater from melting of upland ice. It deposited a great thickness of gravel as a delta into a huge glacial lake (Glacial Lake Penticton) that had formed in the Okanagan Valley. Much of this gravel underlies the Crawford Estates subdivision and has been mined for gravel. When Bellevue Creek began to dissect its present channel a great fan-shaped deposit was formed at a lower elevation. This fan, consisting primarily of boulder gravel, extends all the way to the lake today, and many of the houses and schools of the "Lower Mission" are built on this alluvial fan. Acrylic painting by M. Roed.

APPENDIX A

A SIGHTSEER'S DELIGHT

by Murray A. Roed

The following field trip includes driving directions for a day long trip (100 kilometres) through the Kelowna area focusing on major geologic and landscape features. Reference is made to Chapter Eight, Geologic Landmarks of the Kelowna Area where the features are described in more detail. Each feature is numbered (parenthesis) and located on a map (Figure 1-9) which should be kept handy as the tour progresses.

Start at Knox Mountain (1):

Drive to the top of the mountain by following Ellis Street; look at banded and varved silts of Glacial Lake Penticton on way up (2); park in the parking lot and take the trail to the gazebo view point. Refer to rosette directional map in Figure 2-4.

To Mount Dilworth Drivepast (3):

Go down Ellis, turn left at Clement Avenue, then right at Gordon Drive (site of landslide, all landscaped now, at end of Gordon, see Figure 3-10) and a quick left on High Road. Continue to Mountain Avenue and turn right. Then right on Glenmore Drive. Drive slowly past the Kelowna Golf and Country Club, the cliff bordering the golf course is a steep rock cliff carved out by a glacier. The cliff has been eroded in places into hoodoo landforms (Figures 5-9 and 6-9). The rocks represent feldspathic lava and belong to the Kettle River Group of Eocene age. The base of the cliff can be accessed by a narrow road along the cemetrey.

To Varved Clay Stop on KLO Road (7):

Stay on Glenmore Drive which turns into Spall Road. Go to Springfield Road and turn left. Turn right at the next corner onto Cooper Road then right onto Benvoulin Road. Drive slowly past a small white Church, a local historic landmark (it's open, take a look). Continue to the lights at KLO Road and turn left. After crossing the bridge over Mission Creek and winding up the road, stop at varved clay outcrop on the left(north) side of the road. The road is narrow but there is a small shoulder here. (See Figures 8-9 and 9-9).

To Quarry Stop (11):

Continue east on KLO Road to the viewpoint near the top, pull out and park. This spot offers a good view of valley features, and of dipping rocks in distant Okanagan Mountain Park. Keep going and turn right just past McCulloch Pub onto McCulloch Road. The terrain here is hummocky moraine (8) eroded by a meltwater channel in the vicinity of June Springs Road intersection. Go past a golf course to the bridge over KLO (Canyon) Creek. Drive slowly down the hill past till outcrops in the road cut (10). Park at the pull-out just past the bridge, and walk up the trail to a Quarry in the Shuswap rocks (Monashee Gneiss). This is the same stop as Stop 2, in the Hardrock Trail Field Trip (See Appendix B).

After this, continue a short distance up the hill and watch on the left for a large boulder in the trees; this is an erratic.

To Layer Cake Hill Viewpoint (9):

A viewpoint for Gallagher's Canyon, Pinnacle Rock and Layer Cake Hill at Scenic Canyon Regional Park can be reached by a road through a golf course. Visitors will have to be satisfied with a drive through the new subdivision. This site has received special attention in various parts of this book (refer to Chapter Eight, Figures 11-9, 3-3, 5-4, 12-9, 13-9, 14-9, 15-9, 16-9, and Appendix B). There is a lot to see here.

Crawford Falls (30):

Go back onto McCulloch Road, turn left at June Springs Road and go to West Stewart Road and turn left. Crawford Falls (p.160), can be accessed at the end of Canyon Court in Crawford Estates, East Kelowna. A series of waterfalls occur along Bellevue Creek here but viewing them is presently hazardous. The Kelowna Firestorm of 2003 burned the valley here, and remedial harvesting has substantially improved the views of the waterfalls.

Beach of Glacial Lake Penticton (14):

A spectacular beach ridge can be viewed along the Crawford Escarpment (extension of East Kelowna Escarpment, Figure 1-9) off of Westridge Road (Mission Ridge Park), which joins Crawford Road in Crawford Estates (Figure 1-9). This represents the highest elevation of Glacial Lake Penticton. The beach deposit is up to eight metres thick and overlies sand and gravel of glacial origin exposed in an abandoned gravel pit.

Mt. Boucherie Columns (17):

From Crawford Estates make your way back to Harvey Avenue (Highway 97), perhaps making a stop at the Father Pandosy Historical Site at the corner of Casorso and Benvoulin roads. Also, look at the huge concrete anchor (Figure 1-11) at Anchor Park at the corner of Ellis and Harvey.

Proceed across William R. Bennett Bridge (read the account of this famous bridge on pages 216 - 221). Turn left at the second set of lights onto Boucherie Road, turn right on Hudson Road then left on Guidi Road. Turn right on Trevor Drive and continue to the end of the road where there is a turnaround. Park and walk a short distance. Scattered in the forest here are giant rock columns (Figure 1-A). They are composed of dacite, a volcanic rock that forms part of a volcanic dome that built the mountain. Columnar structures, contorted by folding of the rocks, can be seen high on the east side of the mountain here with binoculars from the turnaround. More detail on Mt. Boucherie is given in Appendix C.

Figure 1-A: Like a field of toppled columns from some ancient Roman ruins, large blocks of columnar lava composed of dacite litter the forest floor on the slopes of Boucherie Mountain. How did these get here, and where did they come from? Photo by Debra Roed.

To Glen Canyon (Powers Creek) (20):

Go back down Trevor Drive, turn left on Guidi Road, right on Hudson Road, then right on Boucherie Road. Stay on Boucherie Road all the way to Gellatly Road. An interesting outcrop (19 in Figure 1-9) of conglomerate of the White Lake Formation occurs at the junction of Old Boucherie Road along this route (Figure 7-B, and Stop 6, the Hardrock Trail, Appendix B).

Turn right at Gellatly Road. Observe varved clay and silt of Glacial Lake Penticton sediments. Turn left at Highway 97 and go through Westbank to Glenrosa Road (Figure 2-A). Turn right here, go to Weber Road and turn right; proceed to Aberdeen Road and turn right. Go to end of Aberdeen and park. There are several viewpoints of a scenic canyon eroded into sandstone and conglomerate of the White Lake Formation at the end of the trail (Figure 7-4). The trail is steep and occasionally in rough condition.

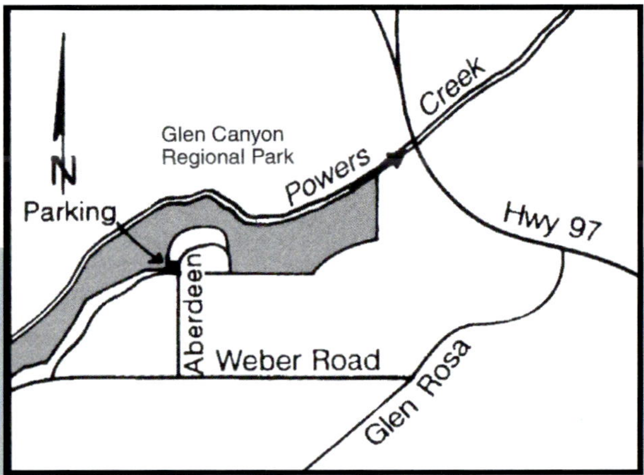

Figure 2-A: Location of Glen Canyon Regional Park.

APPENDIX B

THE HARDROCK TRAIL FIELD TRIP

by John D. Greenough and Leanne Mallory-Greenough

INTRODUCTION

The field trip illustrates processes that occur during mountain building events. The event you will study occurred during the Eocene and strongly affected the Okanagan Valley. Processes illustrated include folding and faulting (stops 1, 4 and 5) igneous activity (stops 1, 3 and 4), metamorphism (stops 2 and 3) and sedimentation (stops 4 and 5). The sedimentary rocks allow us to paint a picture of living conditions in the Okanagan during the Eocene. The trip takes you through East Kelowna, the Okanagan Mission and then along the Kettle Valley Railway bed to Penticton, returning to Kelowna along Highway 97. For those who cannot venture outside of Kelowna stop 3 can be omitted.

STOP 1:
Layer Cake Hill Volcanic Rocks.

Proceed out McCulloch Road, past KLO Creek to where a clearing for high tension power lines provides a spectacular view of layered rocks in Layer Cake Mountain on the north side of Mission Creek (see Figures 10-9, 11-9 and 12-9).

You are standing on Pleistocene sediments that overlie Monashee Gneiss (see Stop 2), and looking at a Middle Eocene dacite (chemically between andesite and rhyolite) lava flow of the Marron Formation in the distance across Mission Creek. This outcrop illustrates that Kelowna was volcanically active approximately 50 million years ago. It also shows that rocks in the area have been folded and faulted since that time.

If you could hold the dacite in your hand you would see that it consists of tiny interlocking crystals, the most obvious being grains of plagioclase feldspar with a few specs of black biotite. The fine-grained, interlocking texture tells you the rock is volcanic. The presence of feldspars and biotite indicate that it is probably a dacite. Formation of the most prominent feature of the outcrop, its layering, has been the subject of much debate.

Many geologists, including the editors initially regarded each layer as a separate lava flow. However, if you look at the outcrop carefully you will see vertical fractures that can be traced across the layers. These fractures are known as columnar jointing, a common feature in volcanic rocks. It forms when initially hot lava cools, crystallizes, and contracts causing the rock to break into regular five- or six-sided polygonal columns. The columns cut across the layering because the entire cliff represents a single cooling unit and is not a series of lava flows.

Little (B.N. Church, personal communication, 1993) recognized that the outcrop is probably a single cooling unit and sarcastically proposed that the inclined layers form the threads on the world's largest "Geotectobolt", a plug in the throat of a volcano that screwed itself to the surface! This thesis seems unlikely because the layers extend back into the cliff. Similar layering occurs in basaltic lava flows. Drill holes through basalts in Hawaii indicate that the layering forms when the crystal-liquid "mush" beneath the hardened upper crust of the lava flow becomes so dense that it begins to sink into the underlying pool of magma. As the "mush" grows downward it tears itself away from the crust at regular time and thickness intervals. The horizontal crack that forms actually sucks liquid out of the crystal mush forming a thin layer of rock between the thick layers. This thin layer is chemically different from the thick layers because the magma that forms it is enriched in the elements not used in forming crystals that were in the mush and now form the thick layers. When the thin layer of magma finally crystallized it formed minerals that are subtly different from those in the thick layers. This hypothesis or model for the formation of Layer Cake Mountain has been tested using chemical data for the rocks (Greenough and Owen, 1998). Curiously, layering like this has never been reported from volcanic rocks of this type anywhere else in the world.

Although you cannot see them, the rocks you are standing on are metamorphic rocks called gneiss that were strongly affected by the Eocene mountain building event. Minerals such as amphibole in the gneiss indicate that the metamorphism (heating) occurred perhaps ten or more kilometres down in the Earth at the same time that Layer Cake Hill was forming at the surface of the Earth. This requires that there is a major Eocene to post-Eocene fault with ten or more kilometres of vertical displacement on it between where you are standing and Layer Cake Mountain! The originally-horizontal layers in Layer Cake Mountain are in some places strongly tilted. Both the folding (bending) and faulting are the result of mountain building forces that acted on the rocks during or after the Eocene.

In summary this stop illustrates three important processes that operate during mountain building events; volcanism, folding and faulting.

STOP 2: Quarry In Metamorphic Rocks - The Monashee Gneiss.

Proceed back toward Kelowna to a quarry located approximately 100 metres south of the road beside KLO Creek (Figures 10-9 and 18-9, Figure 1-B).

Good outcrops of the Monashee Gneiss occur here. Gneisses are metamorphic rocks. By metamorphic we mean "changed in form" by heat and pressure. A geologist tries to answer several questions when examining metamorphic rocks: 1) what was the original (pre-metamorphism) rock type?; 2) how old was the parent rock?; 3) when was it metamorphosed?; 4) what were the pressure and temperature conditions during metamorphism?, and 5) what do mineral alignments in the rock tell us about the orientation of mountain-building forces during metamorphism?

The first thing you would do is identify the minerals and textures in the rock. Minerals present include microcline (K-feldspar), which is white to pink in colour, plagioclase, which is white and very similar in appearance to the microcline, smaller percentages of clear, colourless to light gray quartz, and specs of black minerals which are amphibole and/ or biotite. The mineral grains form an interlocking texture and form light and dark layers. Some sedimentary rocks are layered but those with plagioclase and microcline never show an interlocking texture. Rocks with a banded texture and interlocking crystals are called gneiss.

The minerals present reflect the chemical composition (mineralogy) of the original rock as well as the pressure and temperature of metamorphism. The minerals and mineral proportions are approximately the same as in a granite or its volcanic (extrusive) equivalent, rhyolite. Thus, the gneiss may have originally been a granite. Other possible parent rocks include arkose and greywacke which are quartz-poor and feldspar, or clay-rich sandstones (sedimentary rocks) which form from weathering of granites. Some rocks are better than others for determining the pressure and temperature of metamorphism. The minerals present in this rock are not very good. Nevertheless, the presence of amphibole indicates temperatures of at least 450° C and pressures of 2 kilobars (= depths of at least 7 km) or more. Gneisses of all kinds typically form under these conditions.

Figure 1-B: Photograph of Monashee Gneiss typical of what can be seen in the rock quarry (Stop 2) and at numerous localities along the Kettle Valley Railway bed. The gneiss displays light-colored bands dominated by feldspar and dark bands with abundant amphibole. Augen ("eye" shaped feldspars) in the centre of the photograph have tops that are subtly smeared out to the left whereas the bottoms are smeared to the right. As discussed in the text, these augen provide information on the direction of slip on the Okanagan Valley Fault. Photo by Debra Roed.

The ages of the parent rocks along the fault are variable (e.g. Armstrong et al., 1991). Some rocks belonging to this interior metamorphic complex (see Chapter Two), variably referred to as the Monashee Gneiss Complex, the Shuswap Complex and locally, the Okanagan Gneiss, are approximately two billion years old. Other gneissic rocks along the fault are younger (Paleozoic, Mesozoic and Early Cenozoic). These "protolith" ages are obtained by radiometric dating of zircons that formed at the time of the parent rock and have survived the effects of metamorphism. Zircons occur in such tiny concentrations in the gneiss that they are not usually observable in hand specimen. Radiometric dating of minerals such as amphibole and biotite that formed during metamorphism indicate that the last metamorphic event to affect the rocks occurred during the Eocene. Note that these rocks may have also been metamorphosed many times prior to the Eocene.

The gneisses differ from granites in that they show bands of light and dark minerals. The bands tend to be horizontal. One process that can form banding is shearing. Imagine rubbing (in one direction only) a wad of putty between your hands. The putty becomes smeared out. The same process affected these rocks.

One can determine the sense of shearing by finding eye-shaped crystals of microcline known as augen in some outcrops (Figure 1-B). The augen tend to have "tails" pointing in opposite directions on their tops and bottoms. Relative motion was in the direction of the tails. These crystals indicate that rocks once above your head moved to the west and those below your feet moved east! The boundary between the rocks that moved west and those that went east occurs under Okanagan Lake. You also observed a portion of this fault at stop 1. Thus movement on the Okanagan Valley Fault created the layering now present in the Monashee Gneiss.

STOP 3: Monashee Gneiss (Amphibolite) Intruded by Granite.

From the quarry, return to Kelowna via McCulloch and KLO Roads, turn left on Gordon Drive and right onto Dehart Road and then left onto Lakeshore Drive. Going up the hill leave Lakeshore Drive and follow Chute Lake Road. Watch on the left for the fire station approximately 1.9 km along Chute Lake Road. Turn left off Chute Lake Road onto Hedeman Road and then right onto the Forestry Service road 0.2 km past the fire station. There are no major turns off the forestry service road until you reach the Kettle Valley Railway bed where you turn right and head toward Naramata.

There are numerous railway cuts through bedrock between the logging road-railway bed junction and Naramata all of which contain Monashee Gneiss (Figure 1-B). You are encouraged to stop and compare some of these rocks with those examined at Stop 2. Proceed to the southern end of Penticton and take Lee Avenue off Skaha Lake Road. Head east 0.6 km to South Main Street and turn right. Head south along South Main Street for exactly 5 km and stop on the left-hand corner where rocks on the left have been quarried and used to produce the road bed that abuts Skaha Lake (Figure 2-B). There is lots of room to pull off on the right.

The outcrop illustrates processes that occur during magma intrusion. Granitoid rocks are apparent at the base of the outcrop just before the apex in the corner. These rocks are probably Eocene in age. They intrude

Monashee Gneiss that at this locality is dominantly the metamorphic rock amphibolite as opposed to the "granite gneiss" you saw earlier. The black mineral in the amphibolite is an Iron-rich mineral called hornblende (an amphibole, hence the rock name).

The contact between granite (a light-coloured rock) and amphibolite (a dark rock) is fairly distinct (Figure 3-B). In contrast to the gneisses you observed earlier the amphibolites at this locality are extremely coarse grained. Heat from the granite caused the surrounding gneiss to be metamorphosed and recrystallized. The coarse grained texture in the amphibolite is a result of volatiles leaving the granite. Volatiles, such as water, are light and they tend to rise and be concentrated in the upper portions of granite intrusions. These volatiles, when they entered the gneiss were capable of moving ions to the surface of growing amphibole crystals. Locally you can see veins where the amphibole crystals are extremely large, even "pegmatitic". These represent channel-ways where volatiles were concentrated as they streamed out of the granite. The fluids also enhanced the formation of garnets which occur as abundant, small (< 0.5 cm), round, red mineral grains forming "trails" in the coarse-grained granite commonly about 1 m below the amphibolite (see below for more details on their formation).

Walking south approximately 65 metres around the corner you will find more amphibolites that are not so coarse grained. Close examination shows that some samples also contain a green to dark green mineral which is pyroxene. At this location volatile concentrations were apparently lower and the heat from the intrusion caused amphibole to give up the water in its crystal structure and turn into pyroxene. Very high temperatures, perhaps in excess of 700oC are required for this reaction to take place.

At the same spot there are near-vertical orange to pink granite pegmatites cutting the amphibolites. These may have formed later and resulted from crystallization of the interior of the granite intrusion. Because most minerals in a magma tend to have low volatile contents,

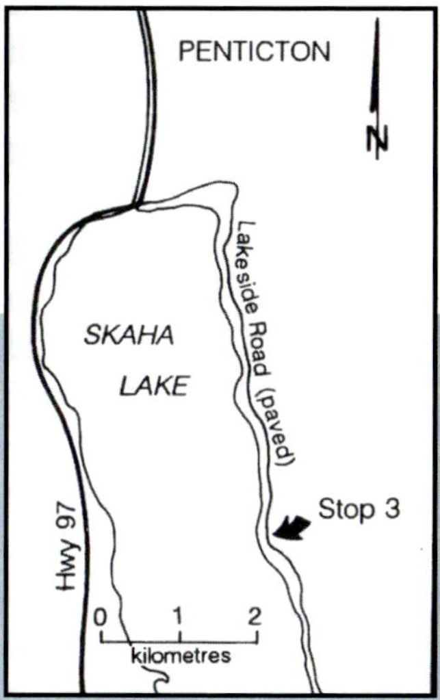

Figure 2-B

Figure 3-B: Photograph showing the sharp contact between granite (bottom of photograph) and coarse-grained amphibolite (top) along Skaha Lake. There is a poorly defined "vein" of extremely coarse amphibolite above the Canadian dollar coin.

Figure 4-B: A granite pegmatite dyke cutting dark amphibolite at the Skaha Lake outcrop. Stop 3.

fluids such as water tend to build up in the residual liquid as the granite crystallizes. Thus, long after the outer edge of the intrusion had completely crystallized, the interior, which solidified more slowly, had volatiles locally build up to the point where they fractured the solid outer margin of the intrusion. Volatile-rich magma then moved along these fractures creating pegmatite dykes. The large grain sizes reflect the ability of volatiles to enhance crystal growth creating large crystals.

It is not only volatiles that build up in the last drops of liquid to crystallize from a magma. Some elements do not fit in the crystal structure of common silicate minerals such as feldspar and quartz. They become concentrated in the last drops of magma to crystallize. Examples of elements that tend to behave this way are zirconium, tantalum, hafnium, and the rare earth elements (e.g. lanthanum, cerium, lutetium). However, some rare minerals in igneous rocks have an affinity for these elements. For example, garnets accommodate the rare earth elements in their crystal structure. Garnet present in the northern portion of the outcrop is probably related to these processes.

STOP 4: White Lake Formation and Coal Deposits, Westbank

Proceed north through Penticton, along Highway 97. You may want to briefly examine the Eocene sediments and pyroclastic rocks in a deep roadcut at the northern edge of Summerland in order to compare them with the rocks in Westbank.

Approximately 1.5 km north of the Okanagan Connector exit ramp there is a place to pull off and examine rocks across the road from the Gorman Brothers Lumber Ltd. mill (Figure 5-B). Walk back (south) 190 m from the parking spot.

Rocks at this stop belong to the White Lake Formation and are Middle Eocene in age (approximately 45 million years old). They are a mixture of volcanic rocks and volcanic-rich sediments that filled a caldera stretching from here to the northern edge of Kelowna. You passed through another one of these caldera-fill Tertiary basins at Summerland. Calderas form as a result of giant volcanic eruptions that leave behind a large hole in the ground. The sedimentary rocks represent rock material that filled the caldera after its formation. They allow us to paint a picture of Okanagan living conditions during the Eocene.

The south end of the road cut is dominated by mudflow deposits composed mostly of coarse-grained conglomerates. Mudflows are characterized by poor sorting. They contain clasts (chunks of older rocks) of many sizes and the clasts sit in a fine-grained matrix and tend to not touch one another.

The clasts in the conglomerate tend to be volcanic rocks. Reddish brown clasts are probably "andesites" because the white specs (crystals) are plagioclase phenocrysts. Other, light-coloured clasts contain K-feldspar phenocrysts and are probably rhyolites. Dark brown clasts with wisps of lighter material in them are ash-fall lapilli tuffs. The 'wisps' of lighter material are lapilli (walnut-sized clots of bubbly magma thrown out of an erupting volcano) that collapsed (became flattened) from the weight of overlying ash. The abundance of volcanic clasts suggests that these are lahars, mudflows that form when snow and ice melt at the top of an erupting volcano. The lahars apparently flowed down a stream valley beyond the base of the volcano because many of the clasts are somewhat rounded and some are granitic (granite-like) and could not have come from a volcano.

Heading north 12 metres, at the upper end of the lahar deposits there is a bed of conglomerate about 1 metre thick with a concentration of large clasts. It represents the top of a lahar that was worked by water flowing down the stream channel. The flowing water removed fine-grained sediment thus increasing the degree of sorting and concentrating the large boulders.

Continuing north 33 metres the sedimentary rocks get finer grained. They are very immature sandstones (greywackes, lithic sandstones and arkoses), and contain abundant volcanic material partially reworked by stream flow. Some beds may even be unworked ash deposits. These rocks give way to thin (< 20 cm) coal beds between ash-fall and/ or sandstone beds. The coal indicates that swamps with abundant plant life periodically developed (Figure 5-B).

Pacing off another 18 metres you will find coalified roots of trees that can be traced down into soils that started to develop in the

Figure 5-B: Bedding in White Lake Formation rocks across from the Gorman Bros. Lumber Ltd. lumber mill at Westbank. Conglomerates discussed in the text begin just out of view on the far right side of the picture and at the southern end of the section. In the middle of the photograph, prominent beds of sandstone are separated by thin, dark coal beds that tend to be eroded more easily. The thick massive bed in the upper left corner is a volcanic ash-fall breccia deposit. The tree trunk and roots occur approximately above Leanne's head.

Figure 6-B: One of several fossil tree trunks with roots (forming an upside down Y) across from the lumber mill. During Eocene time the tree was growing in sand which has been turned into sandstone by the same combination of heat, pressure and time that caused the tree trunk to turn into coal.

volcanic-rich sediments and volcanic ash (Figure 6-B). Fossils of broad-leaf trees, ferns, and forerunners to modern red wood trees indicate a warm and damp, if not sub-tropical climate.

Ten metres further north there are slightly coarser rocks that contain monolithologic (only one rock type, in this case andesite with feldspar phenocrysts) volcanic clasts. This rock is almost certainly a volcanic ash-fall breccia deposit because the matrix (material between clasts) has the same composition as the clasts. Apparently living things in the Kelowna area had a difficult time of it during the Eocene. Floods produced raging rivers that periodically swept away or buried life in swamps. Mudflows (lahars) originating on near-by volcanic edifices buried anything in their path and explosive volcanic eruptions (recorded by the ash beds) regularly smothered everything alive. Despite a pleasant climate, natural hazards meant that this was not a nice place to live!

Beds in the outcrop are not horizontal (Figure 5-B). Sedimentologists generally assume that sediments are deposited horizontally. The beds are no longer flat-lying because they were folded toward the end of the Eocene mountain building event.

STOP 5: Conglomerate Cliff on Boucherie Road

Proceed through the business section of Westbank. Turn right at the intersection with Gellatly Road, and left (head north) onto Boucherie Road upon reaching the lake. Proceed a total of 3.7 km from Highway 97. Watch for a prominent rock cliff containing numerous “holes” on the left-hand side of the road and at the intersection with Old Boucherie Road (Figure 1-9 and Figure 7-B).

These rocks represent more Middle Eocene conglomerate deposits but they are different from those at Stop 4. They show much better sorting (fine grains have been separated from coarse grains by running water) are well-bedded (distinctly layered), and the dominant clast type is granitic. Other clasts of lesser importance include yellow-brown andesite clasts and foliated (mineral grains are lined up) granite gneiss pebbles. The source for these clasts has not been determined but it is likely that they did not come from the immediate Kelowna area. These are river gravel deposits produced by a major river flowing through the caldera complex. One thing these rocks have in common with Stop 4 is the bedding is inclined. This is the result of folding that occurred toward the end of the Eocene mountain building event.

Study of the mineral material cementing the rocks together indicates the origin of the holes in the outcrop. Inspection of freshly blasted rocks at the base of the cliff (sometimes covered by slumping soil) reveals several oval areas where the rocks are lighter in colour. These spots react vigorously with dilute hydrochloric acid (HCl) indicating the presence of calcite as a cementing agent. Away from the oval spots, which are referred to as concretions, the rock does not react with acid.

Apparently soon after the gravels were deposited, groundwater percolating through the sediment deposited calcite at random locations. In the millions of years following this early episode of cementation, quartz dissolved in groundwater cemented the remainder of the rock together. Over the past 10,000 years, rainwater, which is mildly acidic as a result of absorbing C02 from the atmosphere, dissolved away the calcite holding the concretions together. Once the cementing agent was removed the original sand and gravel simply fell out of the rock face leaving the prominent holes.

SUMMARY AND CONCLUSIONS

Mountain-building processes result in the formation of igneous, sedimentary and metamorphic rocks as well as folding and faulting activity. During the Eocene the Kelowna area was affected by a major mountain building event. You saw dacite lava that pooled in the throat of a volcano at the first stop and ash-fall deposits at stop 4. The lahar mudflows at stop 4 can also be regarded as volcanic deposits. Stop 3 illustrated intrusive igneous processes which resulted in metamorphism and recrystallization of amphibolites making up the Monashee Gneiss as well as granite pegmatite formation.

Sedimentary processes were represented at stops 4 and 5. Thin coal seams at stop 4 are products of the accumulation of plant material in swamps and bogs. Conglomerate at stop 5 indicates that a major river flowed through the area dur ing the Eocene. Fossils in the sediments suggest a warm, humid, perhaps subtropical climate. Periodic flash floods, mud flows (lahars) and explosive volcanism made life in the Okanagan difficult during the Eocene.

Metamorphic rocks at stops 2 and 3 contain minerals that formed at depths of perhaps 10 kilometres and temperatures of at least 450°C. These temperatures were the result of regional heating during the

Figure 7-B: White Lake Formation conglomerate showing concretions which have been preferentially weathered (eroded) from the rock face forming "holes".

mountain building event. The foliation in the gneisses (banding and mineral alignment) indicates that rocks that occurred above the gneisses moved toward the west.

Rocks at stops 1, 4 and 5 were folded during mountain building. Faulting occurred, especially toward the end of the event. At stop 1 the juxtaposition of Eocene volcanic rocks (formed at the surface of the earth) and metamorphic rocks (formed simultaneously perhaps 10 or more kilometres down) suggests that 7 kilometres of vertical movement occurred along the Okanagan Valley Fault. The foliation in the Monashee Gneiss indicates that the fault was a low angle fault and that rocks to the west of Okanagan Lake moved dozens if not hundreds of kilometres to the west unearthing the Monashee Gneisses in the east.

All photos in Appendix B were taken by John D. Greenough.

Figure 8-B: Bear Creek Park Viewpoint, acrylic painting by M.Roed This is a view toward the southeast from a viewpoint on the northern side of the creek in Bear Creek Provincial Park. Little White Mountain is in the far distance jutting up from the Aberdeen Plateau. The lowland across the lake is the present site of the City of Kelowna, and the hill to the left is Knox Mountain ending at Poplar Point at the lake edge.

APPENDIX C

GEOLOGY OF BEAR CREEK PROVINCIAL PARK, BOUCHERIE MOUNTAIN & THE KETTLE VALLEY RAILWAY

by Murray A. Roed, John D. Greenough and Brian Hughes

GEOLOGY OF BEAR CREEK CANYON

There is perhaps no better place in the Okanagan to visualise some of the geologic events that have shaped the land than Bear Creek canyon. Most of the major geologic deposits in the canyon can be seen from a high viewpoint on the north side of the valley (Figure 1-C).

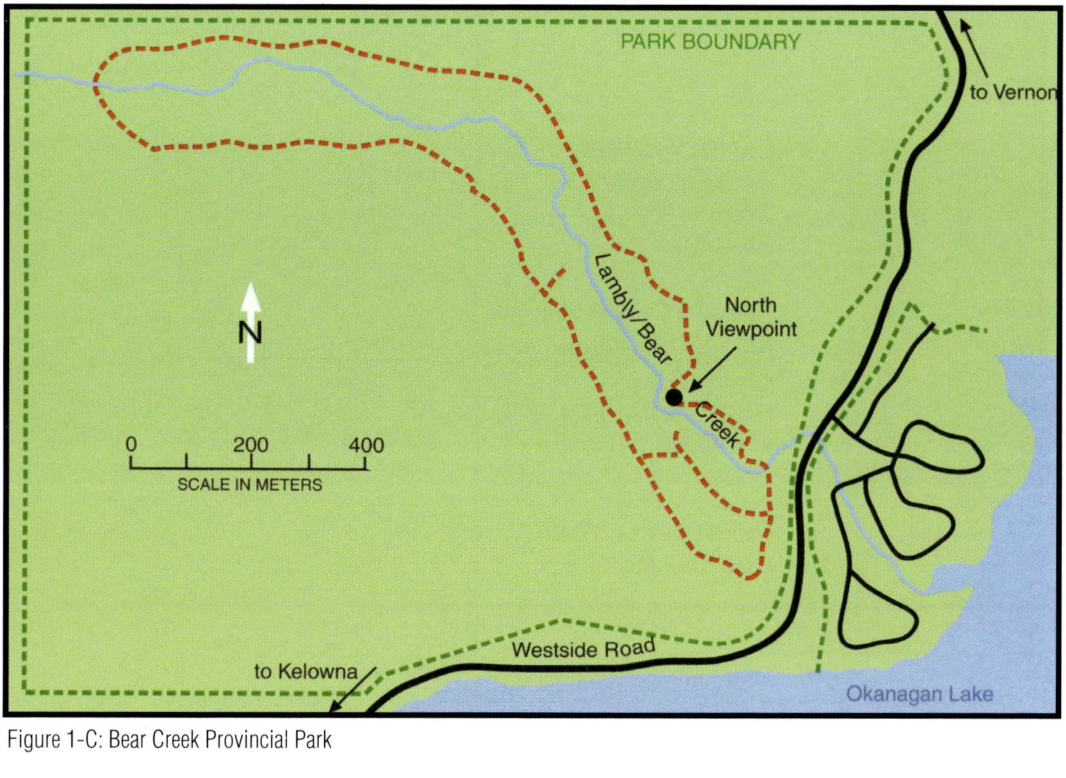

Figure 1-C: Bear Creek Provincial Park

DEPOSITS AND ROCK TYPES

Two major types of geologic material occur in and around the Canyon. One type is unconsolidated and includes sand and gravel deposited by Bear Creek along its stream bed and in a large fan that juts into Okanagan Lake. Other unconsolidated material consists of sand and gravel glacial outwash and till.

The other type of geologic material is solid or consolidated bedrock. There are two types in the area; sedimentary and igneous. Most of the sedimentary rocks are fine grained, grey to black siltstones and sandstones, and minor white siltstones, that have been highly folded and fractured. Igneous rocks include volcanic lava rock (basalt) that poured out onto the surface, and coarsely crystalline granite that formed due to slow cooling of molten rock below the Earth's surface. Thus, both intrusive (granite) and extrusive (basalt) igneous rocks occur in the area.

NORTH VIEWPOINT

Cliffs at the rim of the canyon upland (Figure 8-B) expose glacial till deposited by the Fraser Glacier about 15,000 years ago. Beneath the till is the Lambly Creek basalt (Figure 2-C), locally over 100 metres thick, that erupted from a vent far up the valley 762,000 years ago. It flowed down the valley all the way to the Lakeview area of West Kelowna (Figure 8-4).

Below the basalt cliffs, and challenging to see, is a layer of gravel deposited by an ancient Bear Creek. This gravel may contain the gold found along the present day creek bottom.

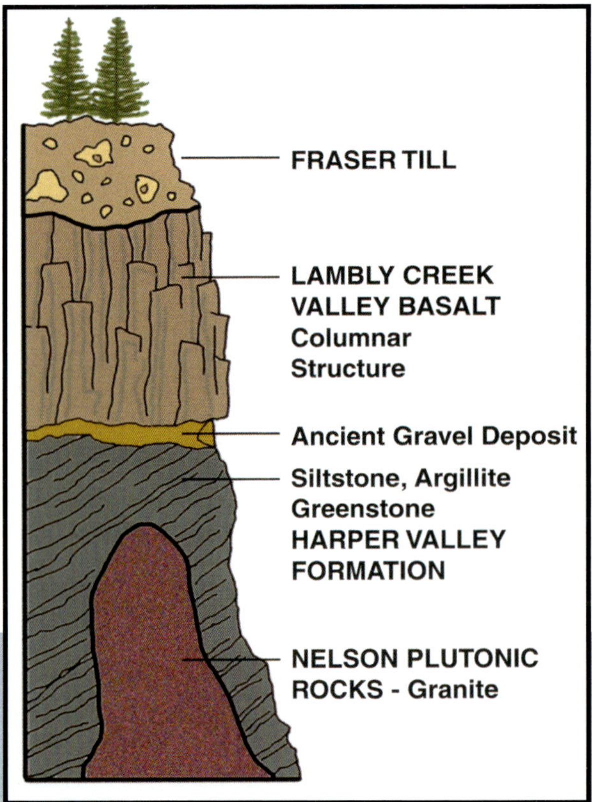

Figure 2-C Representative Geologic Column - Bear Creek Regional Park

An angular unconformity separates the basalt and gravel from an underlying thick sequence of folded and faulted sedimentary rocks associated with the Quesnellia Terrane. Work by Read and Okulitch (1977) suggests these siltstones and other clastic rocks along Bear Creek are mostly Pennsylvanian to Permian in age (now called the Harper Ranch Group). These rocks were originally mapped by Little (1961) who assigned them to the Cache Creek group. As discussed in Chapter 4, a controversy exists whether these rocks, all of which form part of the Quesnellia Terrane, are of exotic origin (e.g. south Pacific) and were added to the continent by plate motions during the Jurassic, or whether they formed on the side of the North American continent and have not been substantially transported. The bald rock cliff forming part of the north side of the canyon is partially of coarse-grained white and pink granite. These igneous rocks belong to the Nelson Plutonic Group of rocks and formed during the Jurassic-Cretaceous Periods. Pieces can be seen strewn along the trail on the north side of the canyon.

WATERFALLS

The Bear Creek falls are spectacular, but like other falls in the Okanagan, it is challenging to get a complete view of them. Water cascades over three main falls along the sinuous and narrow canyon. The overall vertical drop is at least 100 metres, twice as high as Niagara. Obviously the north viewpoint does not depict the falls to full advantage; another view is available on the south trail of the canyon.

DEVELOPMENT OF THE CANYON

The present day Bear Creek canyon was eroded in stages by huge volumes of meltwater derived from the Fraser Glacier, and in a final stage by postglacial erosion The first stage of meltwater, perhaps about 14,000 years ago, formed a channel west of the park boundary that diverted water in a southerly direction along Rose Valley (Figure 3-C). The upstream part of the canyon formed at this time. With continued melting of upland ice and melting of the ice in Lake Okanagan, stage two began

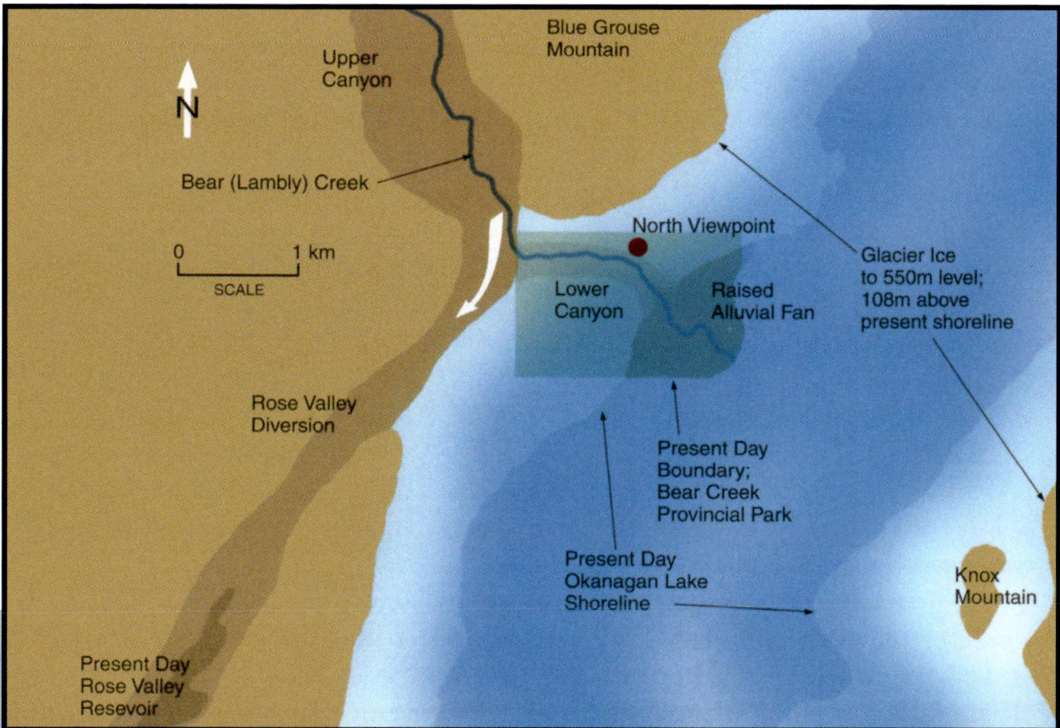

Figure 3-C: Map showing the Fraser Glacier (light blue) in a late stage of melting when the upper canyon at Bear Creek was formed. Meltwater was diverted into Rose Valley at that time.

about 13,000 years ago and water flowed straight into the Okanagan valley. The valley was still largely filled with an ice plug, and water forming Glacial Lake Penticton was at about the level of the north viewpoint. Bear Creek canyon was widened and deepened during this stage. The third stage occurred as Glacial Lake Penticton drained and the present lake dropped to near its present level. Bear Creek cut deeply into the landscape forming a narrow canyon near its mouth. The creek eventually deposited a low alluvial fan or delta at its mouth. This fan now juts into the lake where the campground of the park is located.

GEOLOGIC FEATURES OF MOUNT BOUCHERIE

by Murray A. Roed

Mount Boucherie forms a dominant geologic landmark on the western side of Lake Okanagan. It represents an erosional remnant that formed near the centre of an extensive volcanic complex developed 50 to 60 million years ago, in early Tertiary time. This complex may have stretched from the Kelowna Airport to the Coquihala connector and could have towered 2000 metres high or more. Today, a mere remnant remains but other localities in the Kelowna area where related volcanic rocks are preserved include Knox Mountain, Dilworth Mountain, Layer Cake Hill, Hayman Ridge and Black Knight Mountain. The rocks at Mount Boucherie tell a story about the formation, extinction, burial and final erosion of this Eocene volcano.

CROSS –SECTION

A diagrammatic structural cross-section (Figure 5-C) through Mount Boucherie helps reconstruct the history of the volcano and its caldera. Early major volcanic eruptions witnessed the outpouring of rhyolite and andesite flows that dominate the northern and southern flanks of Boucherie Mountain. These form distinctive yellowish, tan and pinkish rocks. They contain numerous vesicles (cavities) some filled with agate (banded chalcedony, a silicate mineral) and crystalline quartz (see Chapters Two and Twelve). These strata are part of the Marron Formation mapped by B.N.Church, (1980).

COLUMNAR STRUCTURE

Boucherie Mountain is best known for its conspicuous cliffs of black to dark gray dacite that forms spectacular columnar structures (Figure 22-9). These rocks actually represent the last episode of volcanic activity to affect Mount Boucherie as evidenced by the observation that the dacite forms "pipes" that clearly "intrude" or cut through older rhyolite and andesite volcanic rocks. Dacite is a volcanic rock with a chemical composition between rhyolite and andesite. It is extremely fine-grained, so much so that individual crystals are difficult to recognize even with a hand lens, and much of the rock consists of glass. The glass causes the rocks to shatter when hit with a hammer, and produce fragments with conchoidal (smooth and curving) surfaces.

Mapping for this book found that there are several separate dacite pipes on Mount Boucherie (Figure 5-C). One large pipe occurs on the southwestern part of the mountain and another on the northeastern part. Several smaller pipes have been found in the southern and middle portion of the mountain. All of these pipes represent individual "volcanic necks" that were feeders to surface lava "domes", now apparently eroded away. The pipes display intricate, picturesque, contorted columns that may reflect subsurface and near-surface shifts in the orientation of the nearest cooling surface, or movement of near-solid dacite magma just after formation of columns. Large boulders of the columns have locally (e.g. northeast corner) broken away from the cliffs and fallen onto the forest floor below where they mysteriously resemble ruins of ancient temples (Figure 1-A, Appendix A). Although dacite at Mount Boucherie heralds the end of a large volcanoe's active life, there is more to the story. Mount Boucheries is roughly circular with an approximate diameter of two kilometres (Figure 5-C). It now rises only 417 metres above the lake; its highest elevation is 758 metres at the northern edge.

MODIFICATION OF THE ANCIENT VOLCANO

A number of geologic events over the past 40 to 50 million years have greatly subdued the conical form of the mountain from its previous towering grandeur. The first event was the development of a major river represented by the White Lake Formation. This formation has partially buried the roots of volcanoes such as Mount Boucherie, and is now found flanking the base of the mountain. Examples near the mountain include the conspicuous massive cliff of vuggy (holey) conglomerate (Figure 7-B) at the junction of Boucherie Road and Pritchard Drive, and the prominent rock terrace on which Mission Hill Winery is built. Several hoodoos of this rock formation occur on a slope just east of Gregory Road and the Eain Lamont Community Park in Lakeview Place.

Figure 4-C: This is an easterly view of Mount Boucherie taken by John D. Greenough in the spring of 2004.

The second geologic event is related to regional folding and faulting possibly related to the rise of the Rocky Mountains to the East, but also due to regional uplift and mountain building that occurred in the whole Province of British Columbia. The rocks were first squeezed from the west and east. These forces formed northerly trending folds. Then, due to more uplift and major horizontal forces, the rocks were broken up by faults that moved the rock masses up and /or down, and displaced them horizontally.

The last geologic event affecting the mountain involved erosion, first by long intervals of widespread river erosion in the late Tertiary, then by thick glaciers that periodically ground their way across the land during the Ice Age (see Chapter Five). The glaciers molded and streamlined the terrain, preserving the large triangular terrace at Mission Hill Winery that is largely underlain by the White Lake Formation.

A hike to the top of Mount Boucherie, in addition to providing an interesting perspective on the life of a volcano, gives the chance to view a number of important deposits and landforms in the Kelowna area. Major features of the mountain are shown in Figure 6-C.

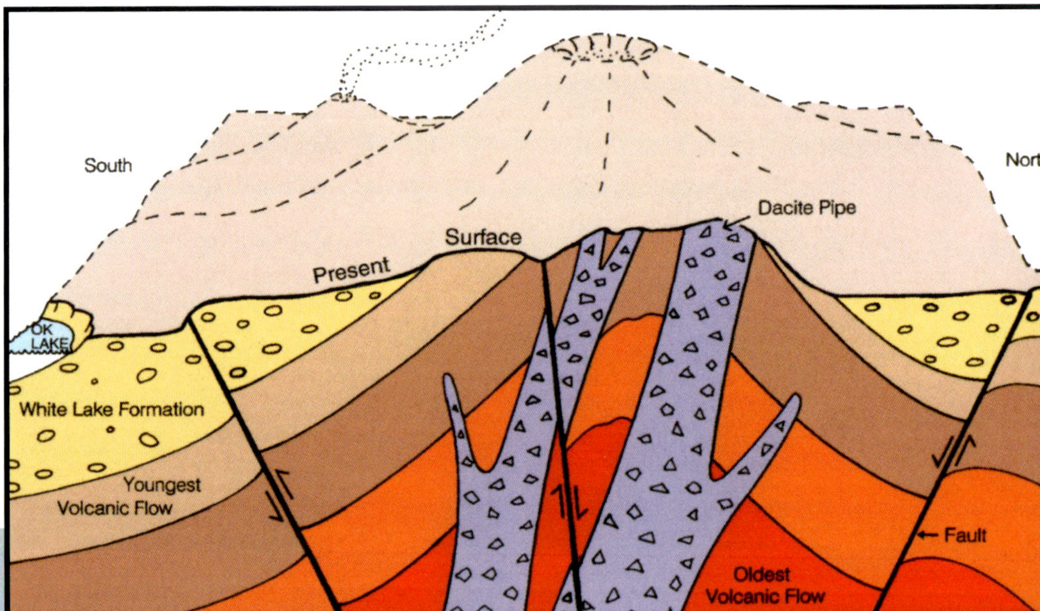

Figure 5-C: DIAGRAMMATIC CROSS-SECTION OF MOUNT BOUCHERIE This is a highly interpretative structural cross-section of the composite volcanic complex of Mount Boucherie. It is not drawn to scale. North is to the right, and South is to the left. The coloured units are deposits from various volcanic eruptions, the last represented by dacite pipes with columnar structure. The near vertical lines represent faults along which there has been differential movement; most of the faults are inferred. Dotted sketches are included in the background to provide an impression of the original volcanic scene in the Okanagan Valley.

MT. BOUCHERIE POINTS OF INTEREST

1. Columnar dacite, dark colour, volcanic necks, northeast and southwest facing cliffs.
2. Andesite, agate infilling some vesicles; vesicular lava of volcano predating dacite volcanic necks.
3. Outcrops of pebble-cobble conglomerate and sandstone, White Lake Formation.
4. Hoodoos eroded from White Lake Formation.
5. Large boulders of dacite with columnar structure; boulders derived from steep east facing cliff and talus slope of Mt. Boucherie.
6. Valley marks the trend of an east-west fault in the volcanic bedrock; movement could be both lateral and up or down.
7. West dipping sandstone and thin coal layers in sediments suspected to be part of the White Lake Formation.
8. Thin bedded rhyolitic volcanic rocks dipping west, part of the older volcanic rocks of the Kelowna Volcanic Complex.
9. Prominent lateral moraine composed of bouldery till was deposited by the Fraser Glacier and encircles the base of Mt. Boucherie on the north, west and east sides.
10. Steep slope exposing abundant columnar structures in dacite volcanic rock which here occurs as a distinct neck or pipe. The volcanic flows appear to be dipping westward.
11. Two small pipes of dacite with columnar structure "Intrude" altered rhyolite-andesite.

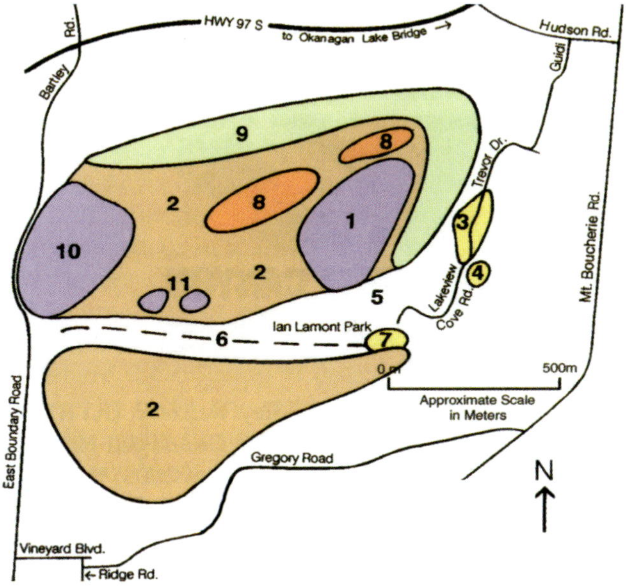

Figure 6-C: Map showing main geologic localities of Mt Boucherie.

GEOLOGY OF MYRA CANYON

by Brian Hughes

The abandoned Kettle Valley Railway (KVR) winds and bends through Myra Canyon Park near the crest of the Okanagan Highland. Despite the loss of most of the trestles (now restored) during the Kelowna fire of 2003, the Myra section of rail bed remains a scenic portion of Canadian history. Along it are excellent exposures of several rock types, metamorphic, sedimentary and volcanic, that are important in the Okanagan and that impacted building of the railway. (Figure 7-C)

THE ROCKS

The most common rocks seen along the KVR are light and dark gray foliated (banded) metamorphic rocks known as gneiss. They are correlated with the Precambrian and younger Monashee Metamorphic Complex (Monashee Gneiss). Examination of the light colored layers reveals white opaque feldspar, with lesser amounts of clear to smoky quartz and black biotite and/or hornblende (an amphibole). Because similar minerals occur in granite, these rocks are sometimes called granite gneiss. The darker layers are composed of white feldspar, black to dark green hornblende, variable amounts of black biotite and minor quartz. Where biotite is abundant one might refer to the rock as a biotite schist because schists are metamorphic rocks where the foliation is defined by the sub-parallel alignment of micaceous minerals. The rocks are amphibolites where mica and

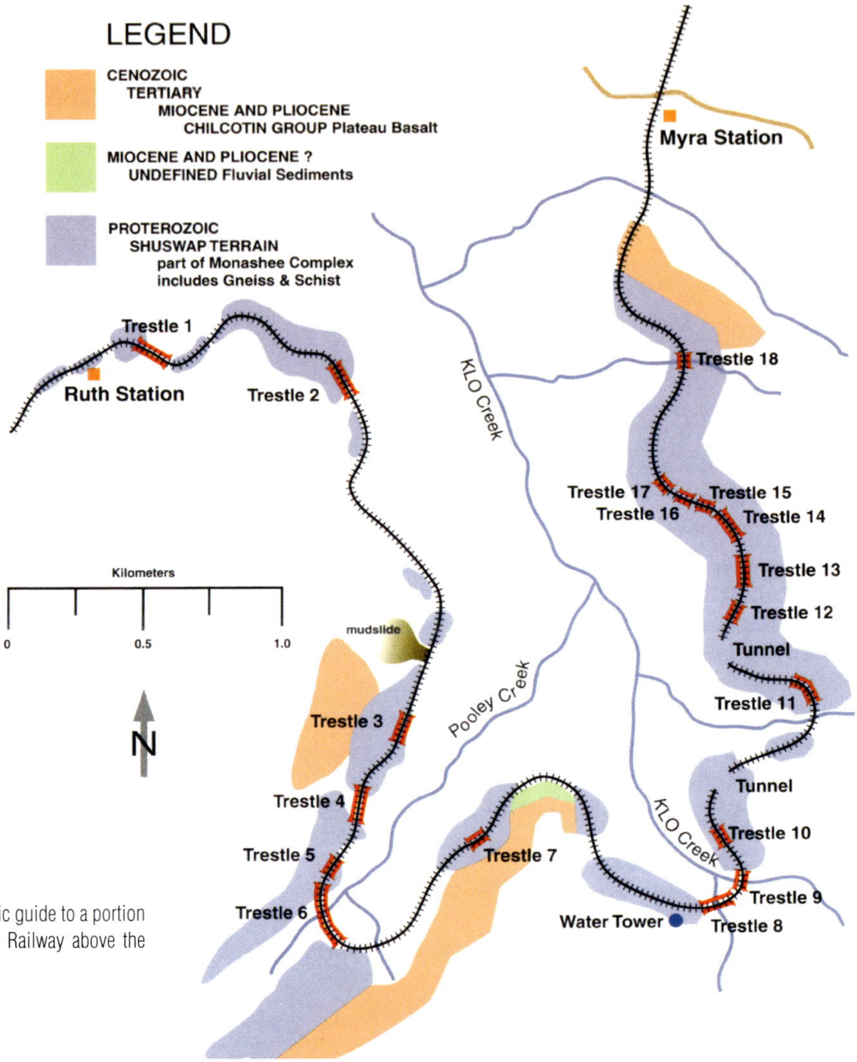

Figure 7-C: Geologic guide to a portion of the Kettle Valley Railway above the City of Kelowna.

a schistosity are less apparent (see Appendix B, stops 1 & 2 for similar rock descriptions). The metamorphic mineral assemblage has probably been overprinted with many metamorphic events, the last being associated with exhumation (bringing to the surface) of Monashee rocks during and following the Eocene orogeny (mountain building 40-50 million years ago).

SEDIMENTARY DEPOSIT

Beds of sandy gravel unconformably overlies the metamorphic rocks and underlies (conformably?) plateau volcanic rocks north of Trestle 7. The precise age of the sediments is not known but many of the plateau basalts in the area are Miocene in age. This basalt-over-sediment environment is host to local uranium deposits as described in Chapter 12.

At Trestle 7 the sediments are composed of poorly consolidated (not cemented together) and poorly sorted (many different fragment sizes, from microscopic to boulders) gravel deposited in a fluvial or stream environment (Figure 9-C). The larger fragments or clasts are metamorphic and igneous in origin with macroscopic portions of the matrix composed of quartz, feldspar and chert grains.

VOLCANIC ROCK

The next most common volcanic rocks are Plateau Basalts (volcanic flows) of Miocene and Pliocene in age. The flows were extruded on an erosion surface over Monashee metamorphic rocks forming an unconformity. This type of contact can be seen between the Myra parking lot and Trestle 18 (Figure 8-C). The basalts were deposited as multiple liquid flows. Flows are separated by disturbed, irregular and vesicular contacts dominated by scoria (vesicular lava) that resembles what the lay-person calls lava-rock. Columnar jointing is evident on cliff exposures of thick flows. The columns form when cooling and contraction of near-solid basalt leads to the creation of six-sided sets of cracks oriented vertical to the flow surface. Portions of these volcanic flows are only 2 to 3 million years old.

Hand specimens of coarse-grained basalt samples (commonly columnar) show minute grains of dark green pyroxene, and locally small olive-green olivine crystals. Rare, large (up to 1 cm), greenish-grey crystals of plagioclase occur in some flows. Vesicles in flow-top samples locally show infillings of quartz, calcite and other minerals that were added after formation of the flow through precipitation from groundwater.

Figure 8-C

ENGINEERING CHALLENGES

The rocks created several challenges for railway bed engineers. North of Trestle 7, ground water percolating down through the overlying columnar jointed basalts has caused slumping and mass wasting (slow down slope movement of rock debris) of the underlying sediments resulting in major rock falls. As overlying basalts are undermined they become unstable and periodically exfoliate (peeling off of layers or

Figure 9-C

slabs) and fall to the cliff base, in this case the railway bed. The wide railway bed north of Trestle 7 is an indication that this has been an on going event.

A second challenge is that low angle (near horizontal) of fracture and foliation dips, combined with steep vertical fractures (joints) in the metamorphic rocks creates unstable, blocky rock faces. In several deep rock cuts made during rail bed construction, the blocky character of the metamorphic rocks has allowed ground water to seep into cracks. Add freeze-thaw cycles and the result is rock falls blocking the rail bed. Several man made solutions are present as concrete and timber retaining structures (Figure 10-C). Similarly, in the tunnels, rock slab detachment along fractured granite gneiss and at amphibolite contacts is evident on tunnel ceilings. The result has been rock slabs falling to the tunnel floors. The impact of rock foliation and fracturing can be seen clearly in the face above the portal cribbing north of where Trestle 11 occurred. Fortunately the rocks seen in Myra Canyon are relatively competent, that is, structurally sound. Thus rock cuts along the railway bed have not collapsed over time, despite irregular maintenance.

CANYON CROSSINGS

Perhaps the greatest challenge was the crossings of deeply incised canyons of Bellevue and KLO Creeks and their tributaries. The canyons were cut into solid bedrock during melting of the massive Fraser Glacier on the Okanagan Highland 13,000 to 14,000 years ago. The canyons themselves are awesome, but what is more profound is left to the imagination of the truly torrentuous volumes of meltwater which must have scoured out this impressive landscape.

Figure 10-C

Appendix D

GEOFACTS OF KELOWNA

1. Maximum elevation of Glacial Lake Penticton is 457.2 metres (1,500 feet).

2. Maximum depth of Okanagan Lake is 232 metres (761 feet) off Whiskey Island, five kilometres north of Okanagan Centre.

3. Maximum thickness of glacial and post-glacial sediment fill in Okanagan Lake is 750 metres (2,460 feet).

4. Elevation of the bedrock floor beneath Okanagan Lake at the deepest is -640 metres (-2100 feet).

5. Okanagan Lake is approximately 120 kilometres long and averages about 3.5 kilometres in width.

6. Elevation of Okanagan Lake is an average of 342 metres (1,123 feet).

7. Maximum water level fluctuation for Okanagan Lake (1943-1979) is from 0 to 1.68 metres (5.5 feet).

8. Oldest rock in the area is included in the Monashee Gneiss, 2.0 billion years or older. Also the oldest rock in British Columbia.

9. Youngest rock in the area is the Lambly Creek Basalt, 0.762 to 1.5 million years, just west of the W.A.C. Bennett Bridge.

10. Highest point of land in Kelowna area: Little White Mountain, elevation 2,171 metres (7,122 feet). terminus of the Crawford Trail.

11. Relief on the bedrock surface in the Okanagan Valley from the highest point (Little White Mountain) to the lowest bedrock beneath the lake is 2,814 metres (9,232 feet), almost double the relief of the Grand Canyon in Arizona at 1,600 metres (5,429 feet).

12. The last glacier that occupied the area is the Fraser Glacier that persisted for 6,000 years from 19,000 to 12,000 years Before Present. This glacier was at least 3,000 metres (10,000 feet) thick along the centre line of Okanagan Lake. An earlier glacier referred to as the Okanagan Centre glaciation occupied the region about 50,000 years ago. There are indirect regional indications that there have been six glacial episodes dating to over one million years ago. A million year old glaciation, the West Kelowna Advance, has recently been discovered.

13. Sixty million years ago, in the early part of the Tertiary Period of geologic time, there was substantial volcanic activity in the Kelowna area, developed just before, during or just after initiation of the Okanagan Valley Fault.

14. The Okanagan Valley Fault, a major break in the earth's crust, has been found to extend to a depth of approximately twenty kilometres (12.4 miles).

15. The Okanagan Valley along which the Okanagan Valley Fault and related structures occur is at least 200 kilometres (124 miles) long in British Columbia and extends far into northern Washington and Oregon, U.S.A.

16. The upland that flanks both sides of the Okanagan Valley is known as the Thompson Plateau on the West and the Okanagan Highland further to the East.

17. The Thompson Plateau and Okanagan Highland are capped in places by nearly flat-lying volcanic rocks of basaltic composition, and range in age from just less than one million years old to twenty million.

18. The Okanagan Lake watershed is 6178 square kilometres (3839 square miles) in area, about one third the size of Switzerland.

APPENDIX E

GLOSSARY OF TERMS

AIR PHOTO: A photograph of the earth's surface taken from the air. It is usually a vertical view, and one of a series of photos taken from an aircraft flying a systematic pattern at a given altitude in order to obtain continuous photo coverage for mapping purposes.

ALLUVIAL FAN: A fan-shaped deposit of fluvial sand and gravel, usually located at the mouth of a tributary valley; a type of floodplain.

ANDESITE: A dark colored, fine grained igneous rock that is the extrusive form of diorite. It is between basalt and rhyolite in chemical composition and was named after the Andes Mountains in South America where it is common.

BASALT: A form of dark-colored lava, composed chiefly of calcium feldspar and pyroxene. Basalt and its coarse-grained equivalent, gabbro, are the principal kinds of rocks that form the crust beneath the deep oceans.

BATHOLITH: (i) A stock-shaped or shield-shaped mass (of igneous rock) intruded as the fusion of older formations. On removal of its rock cover and on continued denudation, this mass holds its diameter or grows broader to unknown depths. (ii) A body of intrusive rock, with the general characteristics of stocks or plutons, but of much larger size than is generally attributed to stocks or bosses.

BEACH: The gently sloping shore of a body of water that is washed by waves and usually composed of loose sandy or gravelly material.

BEDDING: Collective term signifying the existence of beds or laminae. WELL BEDDED indicates beds are immediately apparent, clearly defined and can be easily traced across the deposit; POORLY BEDDED means beds are only discernible after carefully scrutiny, or bedded planes are discontinuous; MODERATELY BEDDED is intermediate between the other two.

BEDROCK: Solid rock, usually older than Quaternary (except rock formed by cooling of lava) either exposed at the land surface or underlying surficial deposits or regolith of varying thickness.

BLACK CHERNOZEMIC: Neutral to basic soils with substantial organic matter accumulation in the surface layer (horizon); black in color.

BLANKET: A mantle of surficial material, thicker than about 1 metre, that reflects the topography of the bedrock or older surficial material upon which it rests although minor details of that topography may be masked.

BOULDER: (i) A rock fragment larger than 256 mm intermediate diameter (Wentworth scale). (ii) Somewhat rounded rock fragment larger than 256 mm intermediate diameter.

BROWN CHERNOZEMIC: Neutral to basic soils with slight organic matter accumulation in the surface layer and developed under climatically dry grassland; brown in color.

BURIED VALLEY: A valley which has been filled by unconsolidated deposits, such as a glacial drift.

CALDERA: A large basin-shaped volcanic depression, more or less circular or cirque-like in form, the diameter of which is many times greater than that of the included volcanic vents or vents, no matter what the steepness of the walls or form of the floor. Classified into three major types: explosion calderas, collapse calderas, and erosion calderas.

CIRQUE: A rounded recess in a mountain formed by glacial erosion, with steep head and side walls, and a relatively gently-sloping floor that is commonly a basin with a small lake and terminated down-valley by a convex break of slope. Cirques range in diameter from a few hundred metres, to several kilometres. They occur at the head of, or cut into the flanks of, glacial troughs. A cirque glacier is a glacier that occupies a cirque.

CLASTIC SEDIMENTS: Sediments consisting of detrital particles derived by mechanical breakdown of rocks.

CLAY: (i) A rock or mineral fragment of any composition having a diameter less than 1/256 mm (4 micrometres) (Wentworth scale). (ii) A finely crystalline hydrous silicate of aluminum, iron, manganese, magnesium, and other metals belonging to the phyllosilicate group, such as kaolinite, montmorillonite, bentonite, and vermiculite, known as CLAY MINERALS.

COBBLE: (i) A rock fragment between 64 and 256 mm intermediate diameter (Wentworth scale). (ii) Rounded and subrounded rock fragments between 62 and 256 mm intermediate diameter.

COLLUVIAL FAN: A fan-shaped mass of sediments deposited by colluvial processes, most commonly debris flows.

CONTACTS (STRATIGRAPHIC): The surfaces that separate a stratigraphic unit from overlying and underlying units; may be sharp or gradational, horizontal or inclined, planar or wavy.

CORDILLERAN ICE SHEET: The complex of ice fields, mountain icecaps and piedmont glaciers that covered much of the Canadian Cordillera, including all of British Columbia, during the Fraser Glaciation for example.

DACITE: A volcanic rock with a finely crystalline or glassy texture, composed of plagioclase feldspar, quartz, pyroxene or hornblende or both with minor biotite and sanidine. It is the extrusive equivalent of quartz diorite.

DARK BROWN CHERNOZEMIC: Neutral to basic soils with moderate organic matter accumulation in the surface layer and developed under grassland; dark brown in color.

DEBRIS AVALANCHE: Rapid downslope movement on steep slopes of saturated soil and/or surficial material, commonly including vegetative debris.

DEBRIS FLOW: Rapid flow of a slurry of saturated debris, including some or all of soil, surficial material, weathered rock, mud, boulders, and vegetative debris. A general designation for all types of rapid downslope flow, including mudflows, rapid earthflows, and debris torrents.

DEBRIS TORRENT: A variety of debris flow that includes little fines (silt and clay) and that follows a pre-existing stream channel.

DEPOSIT: An accumulation of earth material resulting from naturally-occurring physical, chemical, or organic processes.

DELTA: An accumulation of stream-transported sediments deposited where a stream enters a body of water. The landform is flat or very gently sloping, triangular or fan-shaped in plan, and consists of fluvial (alluvial) gravel, sand, silt and/or clay.

DIATOMACEOUS EARTH: A friable earthy deposit composed of nearly pure silica and consisting essentially of the frustules of microscopic plants called diatoms.

DRIFT: (i) All sediments deposited by glacier ice or by glacial meltwater. (ii) Includes till, glacioflluvial and glaciolacustrine materials.

DRUMLIN: A streamlined hill or ridge of till or other drift, with a long axis that parallels the dirction of flow of a former glacier; generally the upstream end is widest and highest, and the drumlin tapers in the downflow direction.

DYSTRIC BRUNISOL: Acidic, reddish to yellowish, strongly weathered soils developed under climatically moist, forested conditions.

EOLIAN MATERIALS: Sediments transported and deposited by wind.

EROSION: The loosening and removal of materials by wind, moving water, weathering, gravity, and glacier ice.

ERRATIC: Boulders or smaller clasts of rock types that are dissimilar to underlying bedrock and transported to their present location by glacier ice.

ESCARPMENT: A steep slope that is usually of great lateral extent compared to its height, such as the risers of river terraces and steep faces associated with stratified rocks or surficial materials.

ESKER: A sinuous ridge of sand and gravel resulting from deposition by meltwater in a tunnel beneath or within a glacier or ice sheet. The ridges generally trend at right angles to a glacier margin, and the sand and gravel may be covered by till or glaciolacustrine sediments.

EUTRIC BRUNISOL: Neutral to basic, weakly weathered soils developed under climatically dry forested conditions.

FAN: (i) An accumulation of detrital material in the shape of a low-angle cone, usually at the point where a stream emerges from a valley onto a plain. (ii) A sector of a cone with a gradient not steeper than 15 degrees. (See alluvial fan, colluvial fan.)

FAULT: Faults are fractures in the earth along which rocks on one side of the fracture have moved relative to rocks on the other side. Faults are further classified on the basis of the kind of movement that has occurred; vertical movement include normal and reverse faults; horizontal movements are strike-slip faults; low angle faults are thrust faults.

FELDSPAR: A group of closely related minerals composed of silica, oxygen, aluminum and one or more of potassium, sodium and calcium. Examples include orthoclase and plagioclase.

FLOODPLAIN: Level or very gently sloping surface bordering a river that has been formed by river erosion and deposition; it is usually subject to flooding, and is underlain by fluvial sediments; similar to alluvial plain.

FLUTINGS: (i) Smooth, straight furrows, parallel to ice-flow direction, and formed in bedrock by glacial abrasion. (ii) Smooth, straight, shallow furrows, parallel to ice-flow direction, in till or other drift.

FLUVIAL: Pertaining to streams and rivers; similar to alluvial.

FOLD: A bend in strata or any planar structure. Upfolds are called anticlines and downfolds are called synclines. Overturned fold is a fold in which one or the other limb has been turned upside down.

FRASER GLACIATION: Name given to the most recent Pleistocene glaciation in British Columbia and adjacent Washington State; equivalent to Late Wisconsinan Glaciation.

GEOLOGICAL STRUCTURE: The three dimensional arrangement of geological contacts and discontinuities, such as bedding, stratification, joints, faults, dikes, plutons and folds.

GEOMORPHOLOGY: The study of the origin of landforms, the processes whereby they are formed, and the materials of which they consist.

GEOMORPHOLOGICAL HISTORY: The evolution of landforms and landscapes, surface materials, and changes with time in geomorphological processes.

GEOMORPHOLOGICAL PROCESSES: Dynamic actions or events that occur at the earth's surface due to application of natural forces resulting from gravity, temperature changes, freezing and thawing, chemical reactions, seismic shaking, and the agencies of wind and moving water, iceand snow. Where and when a force exceeds the strength of the earth material, the material is changed by deformation, translocation, or chemical reactions.

GLACIAL ABRASION: The scouring action of particles embedded in glacier ice.

GLACIAL HISTORY: The time-sequence of glaciations, glacial advances and recessions.

GLACIAL LAKE: (i) A lake that derives much or all of its water from the melting of glacier ice, i.e., fed by meltwater. (ii) A lake that is dammed by a glacier or resting on glacial ice.

GLACIER: A body of ice formed by the compaction and recrystallization of snow, that has definitelateral limits, and with motion in a definite direction.

GLACIER OUTBURST FLOOD: A catastrophic flood that results from the collapse of an ice-dam and rapid drainage of a glacial lake.

GLACIOFLUVIAL: Pertaining to the channelized flow of glacier meltwater (meltwater streams), and deposits and landforms formed by meltwater streams.

GLACIOFLUVIAL MATERIALS: Sediments that exhibit clear evidence of having been deposited by glacial meltwater streams either directly in front of, or in contact with, glacier ice; mostly commonly sands and gravels.

GLACIOLACUSTRINE: Pertaining to glacial lakes.

GLACIOLACUSTRINE MATERIALS: Sediments deposited in or along the margins of glacial lakes; primarily fine sand, silt and clay settled from suspension or from subaqueous gravity flows (turbidity currents), and including coarser sediments (e.g., ice-rafted boulders) released by the melting of floating ice; also includes littoral sediments (e.g., beach sand or gravel) accumulated as a result of wave action.

GLEYSOLIC: Mottled and gleyed soils developed under high water table and saturated conditions.

GNEISS: A foliated, coarse-grained, high-grade metamorphic rock characterized by alternating layers of dark and light-coloured minerals that commonly show flow folding, i.e. typical of rock deformation under plastic conditions.

GRANITE: A light coloured, coarse grained crystalline igneous rock made that is up mostly of plagioclase, potassium feldspar, and quartz; but which may also contain a little mica or hornblende.

GRAVEL: (i) An accumulation of rounded pebbles. (ii) An accumulation of rounded particles that includes at least two of the size classes represented by pebbles, cobbles and boulders; may include interstitial sand.

GRAY LUVISOL: Forested soils with a thick, eluviated (leached) surface layer underlain by a clay accumulation subsurface layer.

GULLY: A small valley or ravine, longer than wide, and typically from a few metres to a few tens of metres across.

GULLY EROSION: Formation of gullies in surficial materials and/or bedrock by a variety of processes including erosion by running water; erosion as a result of weathering and the impact of falling rocks, debris slides, debris flows and other types of mass movement; and erosion by snow avalanches.

HOLOCENE EPOCH: The most recent interval of geological time; from approximately 10,000 years ago to present; similar to postglacial time and also called Recent.

HUMMOCKS: Steep-sided hillocks and hollows, non-linear and chaotically-arranged, and with rounded or irregular cross-profiles; slopes are between 15 and 35 degrees (26-70%) on surficial materials and between 15 and 90 degrees (more than 26%) on bedrock.

HUMMOCKY MORAINE: A moraine consisting of an apparently random assessment of knobs, kettles, hummocks, ridges, and depressions.

HYDROLOGY: The scientific study of the distribution and characteristics of water at and close to the earth's surface.

HYDROLOGIC FEATURES: Refers to water-related features visible at the land surface, such as stream channels, seepage zones, springs, and soil moisture, including soil moisture characteristics as deduced from vegetation characteristics.

HYPSITHERMAL: Early to mid-Holocene warm interval; also referred to as "xerothermic interval" and "climatic optimum"; in British Columbia, warm interval commenced at or shortly after deglaciation; dates reported for end of warm interval are not consistent, and vary from about 6000 to about 2300 years ago.

ICE-DISINTEGRATION MORAINE: A moraine resulting from the accumulation of ablation till and other drift on top of stagnant ice; similar to hummocky moraine and ablation moraine.

ICE-CONTACT: Pertains to sediments deposited against, on top of, or in tunnels underneath a glacier or ice sheet.

ICE-RAFTED STONES: Stones dropped into glaciolacustrine and glaciomarine sediments from melting icebergs; also called dropstones.

ICE SHEET: A continental-scale, more or less continuous cover of land ice that spreads outward in all directions and is not confined by underlying topography.

IGNEOUS ROCKS: Rocks formed from the solidification of molten magma. Those which formed below the earth's surface are coarsely crystalline and are called intrusive rocks. When intrusive rocks are uplifted and uncovered by erosion they are called plutons or batholiths. Igneous rocks that solidified at the earth's surface are called extrusive or volcanic rocks; they are finely crystalline.

INTRUSIVE ROCKS: See igneous rocks.

KAME: Irregular or conical hillocks composed chiefly of sand and gravel; formed by deposition of meltwater-transported sediments in contact with (against, within, or upon) stagnant glacier ice; a type of glaciofluvial deposit.

KAME DELTA: A delta of sand and gravel constructed in contact with (against or on top of) glacier ice; commonly a conspicuous terrace-like landform bounded by a steep ice-contact face or by hummocky collapsed ground; a type of glaciofluvial deposit.

KETTLE: A closed depression or hollow in glacial drift which has resulted from melting of a buried or partly buried mass of glacier ice; common in glaciofluvial deposits.

KETTLE OUTWASH: Outwash plain with kettles.

LACUSTRINE: Pertaining to a lake.

LACUSTRINE MATERIALS: Sediments that have settled from suspension or underwater gravity flows in lakes; also includes littoral sediments (e.g., beach gravels) accumulated as a result of wave action.

LANDFORM: Any physical, recognizable form or feature of the earth's surface, having a characteristic shape, and produced by a natural processes.

LANDSCAPE: A particular part of the earth's surface, such as can be seen from a vantage point or examined on an air photo, and the various landforms and other physical features which together make up the field of view.

LANDSLIDE: A general term for the downslope movement of large masses of earth material and the resulting landforms.

LANDSLIDE SCAR: The part of a slope exposed or visibly modified by detachment and downslope movement of a landslide; usually lies upslope from the displaced landslide material; commonly a steep, concave slope.

LITTLE ICE AGE: The interval of relatively cool/moist climate that occurred during the 15th to 19th centuries during which most mountain glaciers attained their greatest size since the last Pleistocene glaciation; see also Neoglaciation.

LOESS: A homogeneous, nonstratified, not indurated, yellowish to buff-coloured wind borne deposit consisting predominantly of silt-sized particles with subordinate amounts of fine sand, clay; porous and permeable, commonly with incipient vertical joints.

MAGMA: Molten material, mainly derived from the mantle and intruded into the crust, from which igneous rocks are formed.

MASS MOVEMENT: A general term for downslope gravitational movement of earth materials by processes such as rockfall and debris slides.

MASS WASTING: (i) A general term for a variety of processes, including weathering and erosion, that together effect reduction of slopes and lowering of the land surface. (ii) See mass movement.

MASSIVE: Rocks or sediments without stratification, bedding, flow-banding, or foliation.

MATRIX: The groundmass of smaller grains in which larger particles are supported.

MEANDERING CHANNEL: Refers to a stream channel characterized by a series of freely developing curves, bends, loops, turns or windings in the course of a stream.

MIDDEN: A heap or stratum of refuse (broken tools, shells, ashes, etc.) normally found on the site of an ancient settlement.

MORAINE: (i) A landform that consists of till or, less commonly, of other drift; it exhibits a variety of shapes, ranging from plains to mounds and ridges, that are initial constructional forms independent of underlying bedrock or older materials. (ii) See till.

MORAINE RIDGES: Refers to major moraines, such as end moraines, lateral moraines, and recessional moraines, and small moraines, such as washboard moraines.

MORPHOLOGY: The three-dimensional shape or geometry of a landform or other feature.

NEOGLACIAL INTERVAL, NEOGLACIATION: The episode of relatively cool/moist climate during the later part of the Holocene Epoch during which glaciers were more extensive than during the earlier part of the Holocene (see Hypsithermal). In British Columbia, at least three Neoglacial advances have been recognized, approximately 6000-5000, 3400-2200 years B.P. and since about 1000 years ago; the most recent of these intervals is the Little Ice Age.

ORGANIC MATERIALS: Sediments formed by the accumulation of decaying vegetative matter, such as peat.

OROGENY: The process by which structures within fold-belt mountainous areas were formed. Includes thrusting, folding and faulting in the outer higher layers, and plastic flow, metamorphism and plutonism in the inner and deeper layers.

OUTWASH: Glaciofluvial sediments deposited by glacial meltwater downstream from a glacier.

OUTWASH PLAIN: A flat or very gently sloping surface underlain by glaciofluvial sediments.

PALEOZOIC ERA: The span of geological time from 570 to 230 million years ago, i.e. the beginning of the Cambrian to the end of the Permian Periods.

PEBBLE: (i) A rock fragment between 2 and 64mm intermediate diameter; (Wentworth scale). (ii) A rounded rock fragment between 2 and 64mm diameter.

PEDOLOGICAL: Pertaining to the study of the formation and development of soil at the land surface.

PERMEABLE: A material through which water can pass.

PHYSIOGRAPHY: Pertains to the factors that influence the development of landforms or a landscape, such as relief or topography, bedrock geology and structure, and geomorphological history.

PHYSIOGRAPHIC REGION: An area of similar relief and topography, bedrock geology and structure, geomorphological history and landforms.

PIPING: Subsurface erosion of particulate materials by flowing water, resulting in the formation of underground caves and conduits and the development of collapse-depressions at the land surface.

PITTED OUTWASH (PLAIN): An outwash plain with kettles.

PLACER DEPOSIT: A surficial mineral deposit formed by mechanical concentration of mineral particles from weathered debris. Common types are beach placers and stream placers. Heavy minerals are usually concentrated, like magnetite, limonite, garnet, gold, rutile, platinum and cassiterite.

PLAIN: (i) A level or very gently sloping planar surface with gradient up to 3 degrees (5%); local relief is less than 1 m. (ii) An extensive region of comparatively smooth and level or gently undulating land, having few or no prominent surface irregularities, and usually at a low elevation with reference to surrounding areas.

PLATE TECTONICS: A theory of global tectonics in which the lithosphere is divided into a number of plates whose pattern of horizontal movement is that of torsionally rigid bodies that interact with one another at their boundaries, causing seismic and tectonic activity along these boundaries.

PLEISTOCENE: An epoch of the Quaternary Period, after the Pliocene of the Tertiary and before the Holocene; characterized by repeated glacial and non-glacial intervals; also, the corresponding worldwide series of deposits.

PLUTON: Pertaining to a body of intrusive and igneous origin, commonly granite or granodiorite.

PODZOLIC: Reddish and brownish forested soils developed under cool and moist conditions with subsurface layers enriched with iron and aluminum sesquioxides and organic matter. A layer of forest litter is usually present on the soil surface and is often underlain by a thin, whitish leached mineral layer.

PORE SPACES: The spaces between the particles of detrital sediments that are not occupied by mineral matter.

POROSITY: The amount of pore space present, expressed as a percentage of the total volume of the material.

POSTGLACIAL: Pertaining to the time interval since the disappearance of glaciers or an ice sheet from a particular area; similar to Holocene Epoch.

PRECAMBRIAN: The span of time from the origin of the Earth to the beginning of the Cambrian Period; i.e. about 4,000 million years.

PYROCLASTIC SEDIMENTS (MATERIALS): A general term applied to detrital volcanic materials that have been explosively or aerially ejected from a volcanic vent, such as ash and cinders.

QUATERNARY PERIOD: The most recent geological time period; subdivided into the Pleistocene and Holocene (Recent) Epochs; currently defined as beginning about 1.6 million years ago.

QUATERNARY DEPOSITS (MATERIALS): Sediments deposited during the Quaternary Period; similar to surficial materials.

RADIOMETRIC (AGE) DATING: A method of determining the age of rocks and minerals by measuring the ratio of radioactive parent elements and the stable daughter elements into which they decay.

REGOSOLIC: Young soils which do not exhibit soil layers typical of soil formation.

RICHTER SCALE: Refers to an earthquake magnitude scale based on the response to ground motion of a "standard seismograph" located 100 kilometres from an epicentre. The scale is logarithmic, meaning that the recorded amplitude of a magnitude 7.0 earthquake is ten times stronger than a magnitude 6.0 earthquake, or hundred times stronger than a magnitude 5.0, and so on.

RIFT VALLEY: An elongated valley formed by the depression of a block of the earth's crust between two faults or fault zones of approximately parallel strike, a graben.

RIVER TERRACE: A more or less flat surface bounded downslope by a scarp and resulting from fluvial erosion and deposition. Same as fluvial terraces and alluvial terraces.

ROCHE MOUTONNÉE: A knob of rock with a whale-back form, the long axis of which is oriented parallel to former ice flow, and having a smooth, glacially-abraded stoss (up-flow) slope and a much steeper and rougher, glacially-plucked lee slope.

ROCK SLIDE: Rapid or slow downslope movement of a large mass of rock by sliding along one or more well defined surfaces of rupture.

SAND: A detrital particle having a diameter in the range of 1/16 to 2 mm.

SEDIMENTARY ROCKS: Rocks resulting from the consolidation of loose sediment that has accumulated in layers, or strata. The sediments may be produced by mechanical means (fragments of older rocks transported from a source and deposited in water or from air), chemical means (precipitates from solution) or organic means (e.g. limestones constructed from the remains or secretions of plants or animals).

SEEPAGE ZONE: An area where soil is saturated due to emerging ground water.

SEISMIC: Pertaining to earthquakes.

SILT: A detrital particle having a diameter in range of 1/256 to 1/16 mm (0.004 to 0.0625 mm).

SINKHOLES: A funnel shaped depression in the land surface that communicates with a subterranean passage developed by solution; common in limestone and karst regions: also applied to similar features caused by piping.

SLOPE FAILURE: Rupture and collapse, or flow, of surficial materials, soils, or bedrock due to shear stress exceeding shear strength or the material.

SLOPE STABILITY: Pertains to the susceptibility of slope to landslides and the likelihood of slope failure.

SLUMP: A landslide characterized by a shearing and rotary movement of a cohesive mass of rock or earth along a concave, upwardly curved slip surface; rotational slump.

SLUMP-EARTHFLOW: A complex landslide displaying characteristics of a slump in its headward zone, and characteristics of an earthflow in its downslope zone.

SLUMP STRUCTURE: Warped or faulted bedding or stratification within a deposit, resulting from downslope movement due to gravity since deposition.

SOIL: The natural medium for growth of land plants; the result of the combined effects of physical, chemical and biological processes.

SOIL (ENGINEERING): See surficial materials, Quaternary materials.

SOIL DRAINAGE: Refers to the rapidity and extent of water removal from the soil in relation to additions, especially by surface runoff and by percolation downwards through the soil.

SOLONETZIC: Soils containing high proportions of sodium, magnesium and other salts.

STAGNANT ICE: Part of the glacier or ice sheet within which ice is no longer flowing; stationary ice; usually melting by down wasting.

STEEP SLOPE: A planar surface steeper than about 35 degrees (70%).

STEREOPAIR: Two adjacent photos from a fight line; can be viewed simultaneously under a stereoscope to obtain a three-dimensional image.

STRIAE, STRIATIONS: Fine cut lines (scratches) on the surface of bedrock or clasts formed by glacial abrasion; oriented parallel to former ice-flow direction; more than one ice-flow directions may be represented by criss-crossing striae.

SUBDUCTION: A process of one lithospheric plate descending beneath another. A related concept was originally used by Alpine geologists.

SURFACE EXPRESSION: Refers to small topographic features and landforms that are not usually shown adequately on a topographic map, and to the relation of a surficial material to the underlying surface; terminology, such as "terrace", "cone", is defined in a non-genetic sense.

SURFICIAL DEPOSITS (MATERIAL): Relatively young, non-lithified sediments, usually of Quaternary age; usually classified as to their genesis, hence fluvial sediments, colluvium, glaciolacustrine sediments, etc.

SURFICIAL GEOLOGY: Geology of surficial deposits.

SURFICIAL GEOLOGY MAP: A map that shows the types and distribution of surficial materials in a chronostratigraphic framework.

TALUS: Angular rock fragments accumulated at the foot of a steep rock slope and being the product of successive rock falls; a type of colluvium.

TERRACE: Any relatively level or gently inclined surface, generally less broad than a plain, and bounded along one side by a steeper descending slope or scarp and along the other by a steeper ascending slope or scarp.

TERRAIN: (i) A comprehensive term to describe a tract of landscape being studied with respect to its natural features. (ii) Pertains to maps showing the origin of surficial material, material texture, surface expression, present day geomorphological (geological) processes, and related features.

TERRAIN ANALYSIS: The process of terrain mapping and interpretation or assessment of terrain conditions for a specific purpose such as construction of logging roads or urban expansion.

TERRANE: Area of lithosphere distinguished by a certain assemblage of rock types.

TILL: Material deposited by glaciers and ice sheets without modification by any other agent of transportation. Includes basal till, lodgement till, ablation till, flow till, supraglacial till, meltout till.

TILL PLAIN: A level or gently undulating surface underlain by till.

UNCONFORMITY: A substantial break or gap in the geologic record where a rock unit is overlain by another that is not next in stratigraphic succession, such as an interruption in the continuity of a depositional sequence of sedimentary rocks or a break between eroded igneous rocks and younger sedimentary rocks.

VALLEY FILL: Surficial materials that fill or partly fill a valley.

VALLEY GLACIER: Glacier confined by valley sides; usually much longer than broad.

VARVES: Sedimentary beds or laminae where annual layers are distinguishable; most commonly present in glaciolacustrine and lacustrine sediments.

VENEER: A thin mantle of surficial material that does not mask the topographic irregularities of the surface upon which it rests; ranges in thickness from 10 cm to about 1 m.

VOLCANIC ASH: Pyroclastic material less than 2 mm in size.

WATER TABLE: The upper surface of the zone of groundwater saturation in permeable rocks or surficial materials.

WEATHERING: The decomposition and disintegration of bedrock in situ due to chemical and physical processes

Figure 1-E: Sunset behind Mt. Boucherie mimics a volcanic eruption. View to the South at Manhattan Beach.

APPENDIX F

INDEX

Lower Crawford Falls, acrylic painting by M.Roed

Hoodoos developed in silt deposits of Glacial Lake Penticton overlook Okanagan Lake at Sage Mesa in Penticton. This acrylic painting by M. Roed depicts the valley prior to modern residential development. Part of Munson Hill is shown on the left across the lake and the silt bluffs of Naramata exposed on the far shore of the lake. The valley in the background is Penticton Creek.

ADDENDUM

NEW GEOLOGIC DISCOVERIES AND PROJECTS IN THE OKANAGAN

by Murray A. Roed

Science does not stand still. The first book on Okanagan geology appeared in 1995 (Roed, 1995). This book was considerably enlarged with a second edition in 2004 (Roed and Greenough, editors, 2004). Then another book on "Okanagan Geology South" (Roed and Fulton, 2011) reported numerous new geologic features and presented reconnaissance geology maps of all of the south Okanagan, including town sites. Both bedrock and surficial deposits are shown on the maps.

The latest geologic discoveries reported in this addendum include: (1) evidence of a major glaciation over one million years old that affected not only the Okanagan but all of the Interior of British Columbia (Roed et al. 2014); (2) discovery of a variety of spectacular bottom features beneath Okanagan Lake revealed during subaqueous sonar surveys conducted by Gene and Sandy Ralston since 2009; (3) a new bedrock coloured compilation map for all of the Okanagan Watershed published by the Geological Survey of Canada in 2013 (Okulitch, 2013); (4) a new coloured surficial geology map of the Kelowna area; and (5) of geologic and engineering interest, the new William. R. Bennett Bridge was completed in 2008 utilizing some remarkable innovations reported here (Uthayakamur and Naesgaard, 2011). This 3rd edition of Okanagan Geology thus presents the stories concerning these interesting new discoveries and projects in this Addendum.

EVIDENCE FOR AN EARLY PLEISTOCENE GLACIATION IN THE OKANAGAN VALLEY

A multi-disciplinary study of glacial deposits discovered beneath a valley basalt during construction of the Highway 97 interchange in West Kelowna, British Columbia, resulted in defining a regional glaciation called the West Kelowna Advance that occurred at least one million years ago (Roed et al, 2014). The basalt is referred to as the Lambly Creek Basalt. A series of three valley lava flows that erupted near the headwaters of Bear (Lambly) Creek, flowed down the creek and turned south in the valley. The flows continued to

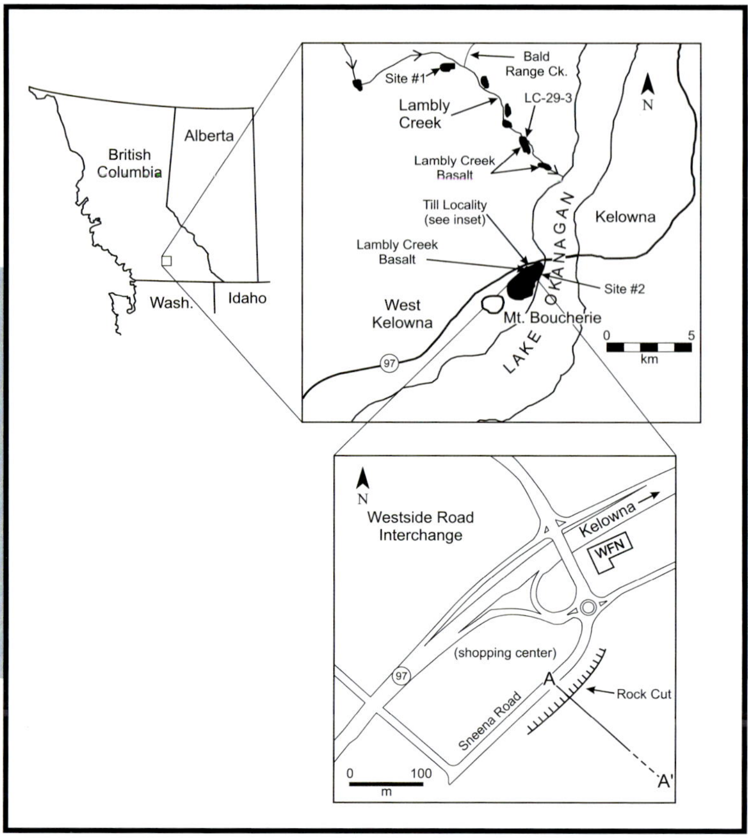

Figure 1-AD: These maps show the occurrence of the Lambly Creek Basalt along Lambly (Bear) Creek and beneath Lakeview Heights. A detailed inset shows the till discovery locality in a rock cut along Sneena Road. WFN refers to Westbank First Nation building.

near the base of Mount Boucherie (Figure 1-AD) covering up old glacial deposits beneath Lakeview Heights residential area.

Remnant basalt outcrops occur along an east-facing scarp of Lakeview Heights in West Kelowna and in a rock-cut at the study site along Sneena Road (Figure 1-AD). Five samples of the basalt provided new 40Ar/39Ar dates (radiometric dates) with a weighted average age of 0.95 ± 0.05 Ma (million years). Beneath the basalt, glacial deposits, from oldest to youngest, consist of a dark grey banded clay overlain by layered gravel, in turn overlain by fluvial (river) and lacustrine (lake) sand. These sediments are all indicative of deposits laid down in front of an advancing glacier. A dense, massive till up to 3 metres thick overlies this sequence and underlies the basalt (Figure 2-AD). The till, referred to as the Westbank First Nation till, contains clasts originating from a glacier that eroded granitic and volcanic terrain on the Thompson Plateau to the northwest of the study site. Fabric analysis (compass direction of the axis and slope of long pebbles) of clasts also supports this source, and suggests that the Okanagan Valley had little influence on flow direction of this southeasterly moving regional ice dome. Striated volcanic clasts were observed but are not numerous. A brunisolic paleosol (ancient soil) occurs on the till suggesting a semiarid grassland ancient climate prior to extrusion of the valley basalt flow, similar to our present climate. Pollen apparently did not survive the basalt flow and prolonged burial. The main deposits at the rock-cut study site are shown in the columnar section of Figure 3-AD.

Paleomagnetic measurements were conducted on all sediments (including till and paleosol) and the basalt to further constrain the age of these deposits. Magnetic particles get "frozen" when deposits form and align to the magnetic North-South pole (varies through time) at the time of deposition. World-wide studies like this by numerous scientists have allowed development of a detailed geologic time chart useful for dating deposits. All samples at the study site are normally magnetized and individual basalt flows appear to have been deposited over a relatively short period of time, based on analyzed data. Both sediments and basalt fall within what is known as the Jaramillo subchron (1.069-0.987 Ma), a part of a paleomagnetic time chart. This is the only significant cold temperature peak within the subchron.

A review of other Early Pleistocene glaciation localities has revealed that central British Columbia was the site of a major glaciation at least one million years ago. The West Kelowna Advance documented at the study site provides the first proof that this glaciation was regional in scope, and impacted the Okanagan Valley with an ice thickness on the order of 3000 metres. It probably took three to four thousand years for the West Kelowna Advance to melt. During this time the valley was deeply eroded by meltwater and later river action. Thus, when the Lambly Creek Basalt flows occurred, there was a substantial basin formed into which the flows were deposited (Figure 4-AD). The flows covered up and preserved the underlying old glacial deposits found at the rock-cut study site. A happy thing for the ten scientists involved in this research.

Figure 2-AD: This is a northwest view of the glacial deposits beneath the Middle flow of the Lambly Creek Basalt at the rock cut locality. The Westbank First Nation Till is the grey sediment immediately under the basalt. The sand below the till has been disturbed by the passing of the glacier, and underlying clay and plugs of gravel have been injected into the sand; minor faults also affect the sand unit.

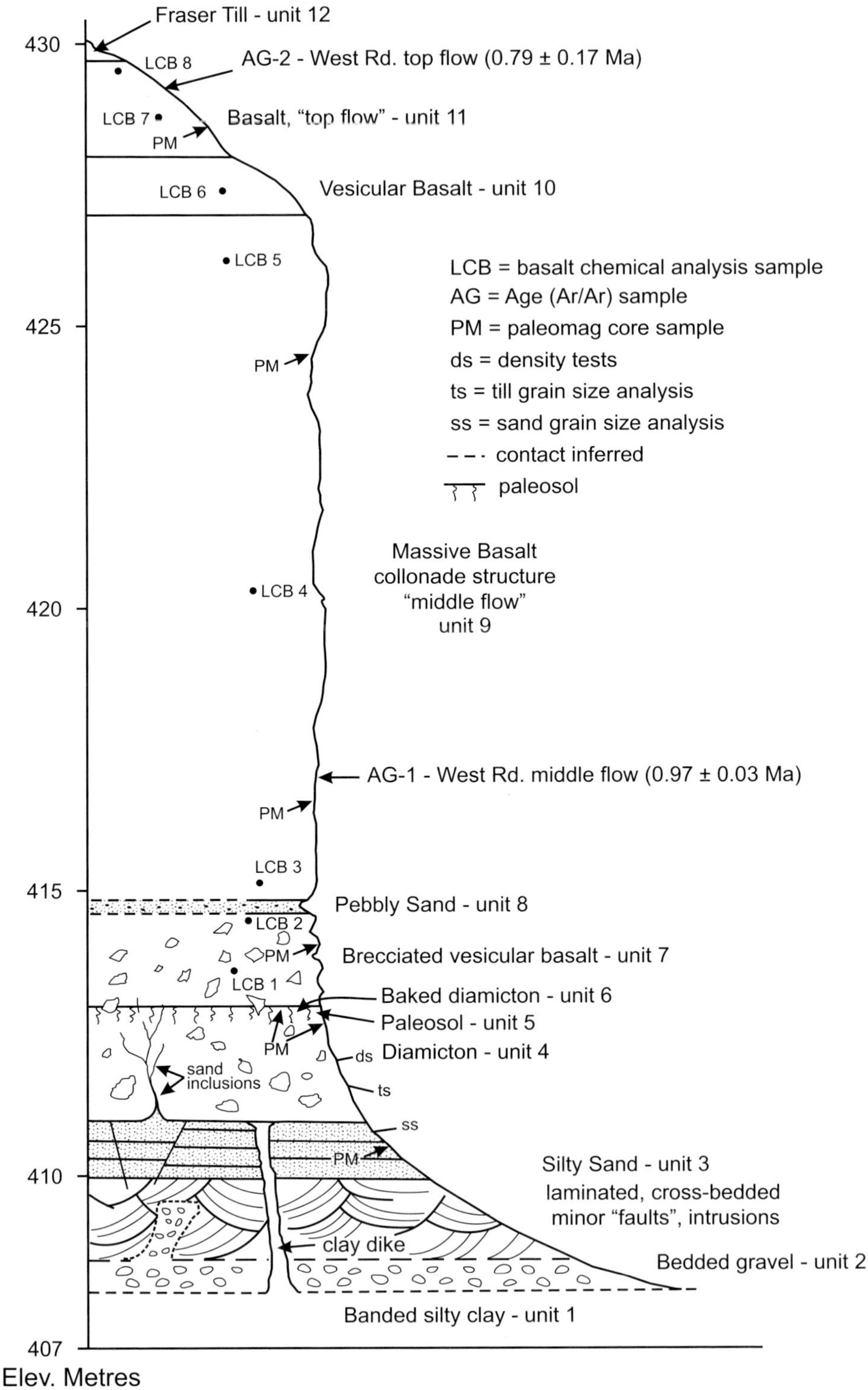

Figure 3-AD: Detail of the stratigraphy at the rock cut is shown here. Symbols refer to actual sample sites for the various tests that were done. (e.g. AG - age dating sample, DS - density test, etc.). Diamicton is till.

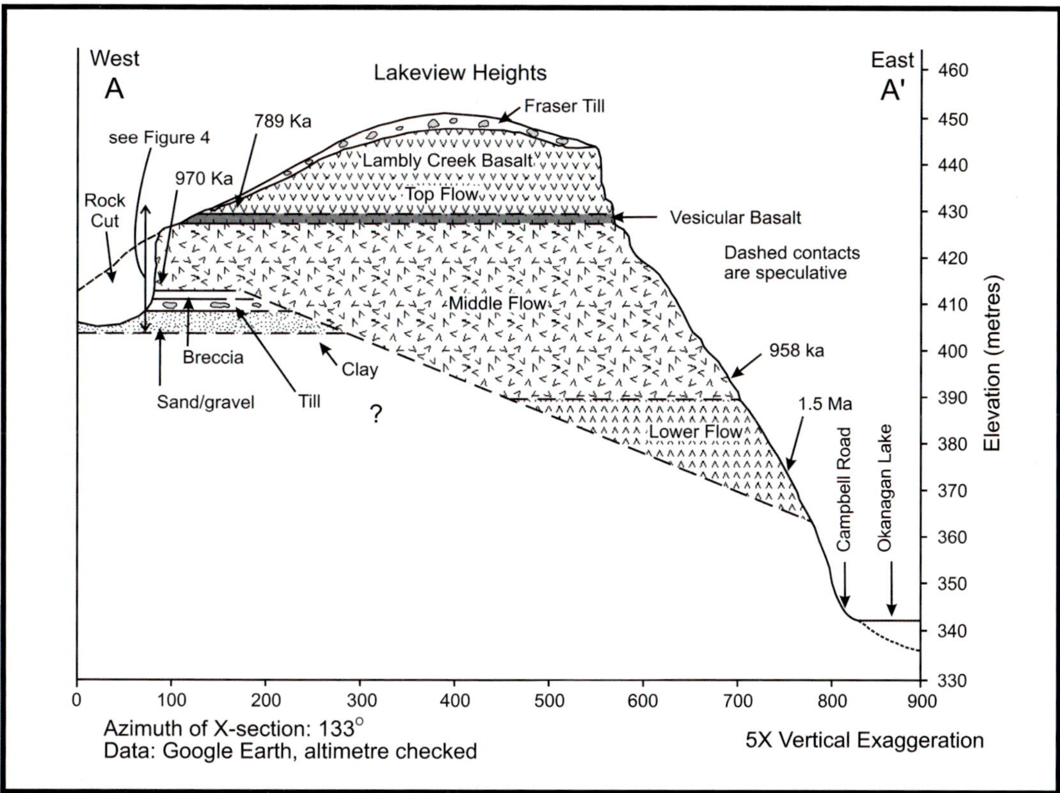

Figure 4-AD: This is an west-east cross-section through Lakeview Heights showing the glacial sediments beneath the middle flow of the Lambly Creek Basalt and location of the rock cut. Numbers like 958Ka are radiometric age determinations, i.e. 958,000 years.

Bottom Features of Okanagan Lake

Numerous gently sloping to flat plains on the bottom of a small portion of Okanagan Lake (Figure 5-AD) reveal a fantastic array of geologic features in water depths from 50 to 120 metres. These features were first detected by Gene and Sandy Ralston during deployment of their sonor-equipped motor vessel during surveys for drowned victims in the lake since 2009. The features consist of "pictures" of a wide variety of reticulated low relief plains marked by pits and furrows up to one metre in width and a half a metre deep, and bizarre patterned areas larger than 30 metres wide in the form of rills and coalesced pits that resemble a modern art work (playfully dubbed "Ogopogo art"). Several subaqeous terrain types are shown in Figure 6-AD which is a mosaic of a number of side-scanning photos prepared by Geof Mullins (JKL RESEARCH Ltd.); the location of this and other mosaics are given in Figure 5-AD. The area surveyed is far less than 1% of the area of Okanagan Lake.

ORIGIN OF THE BOTTOM FEATURES

These features formed on the bottom of the lake since melting of the Fraser Glacier about 12,000 years ago. Sediment from various sources has been deposited on top of the glacial deposits since then. These include sediment from Glacial Lake Penticton, from present streams entering the lake, and sediment from atmospheric sources such as volcanic ash and dust storms. However, the valley was partly filled with older glacial, fluvial and lacustrine sedimentation during at least two previous glaciations and interglacial periods, and it is, in fact, a buried valley (Figure 7-AD). The maximum thickness of all these sediments is 750 metres (2,460 feet). Very limited sampling of the surface of the lake bottom has been done in the vicinity of the bottom features; all are composed of a soft, highly plastic gray clay of unknown thickness.

GEOLOGIC PROCESSES AT LAKE BOTTOM

Consideration of potential geologic processes acting in or on the lake bottom sediments include hydrostatic groundwater pressure (springs), organic decay creating natural gas within the sediments, and disturbance from minor earthquakes. Groundwater movement in bedrock from areas of high relief is discharged at lower elevations as springs. Uplands in the Okanagan are approximately 1500 metres above sea level and the bedrock below the lake is 1200 metres **below** sea level. This adds up to a relief of 2700 metres (over 8800feet) for the bedrock surface and creates a substantial artesian condition along the length of the lake. So it is possible that some of the bottom features are caused by groundwater "bubbling" up through sediments at the lake bottom and forming, for example, the pits. The same kind of feature could be caused by seeps of natural gas that originate from organic decay in the very thick section of older sediments in the buried valley (Figure 7-AD). Even the more bizarre bottom features ("Ogopogo art") could be formed by springs or gas discharge in broad zones, but little is known about such processes. Minor earthquakes occur in the Okanagan on a yearly basis and presumably they occurred regularly since at least the end of glaciation 12,000 years ago. The bottom features are developed in an upper mantle of the soft clay, of unknown thickness. This clay is considered very sensitive and if disturbed could even flow or slide out of place at a very low gradient of say, a one degree slope. Amongst the bottom features are dislocations (faults) of patterned features, prominent "scars" related to some kind of bottom movement, and pockets of linear structures that look like slumps or slides (Figure 8-AD). All of these features could be caused by the effects of minor earthquake energy in low-sloping subaqueous plains. There are presently no definitive answers concerning the origin of any of these bottom features of Okanagan Lake.

Bedrock Geology of the Okanagan Watershed

A portion of a new compiled bedrock geology map of the entire Okanagan Watershed (Okulitch, 2013) is shown for the Kelowna area in Figure 9-AD (compare this with an old version shown in Figure 1-4). The present compilation spans maps of the region going back 70 years. The compilation was challenging to create due to the complex and diverse sources of data, and because the bedrock types are shown as if there were no surficial deposits that, in fact, mask much of the rock terrain. Despite many studies there is still a lack of understanding of the numerous map units and not nearly enough age determinations to confidently offer an advanced concept of the geological evolution of the region. The following is a very brief summary of the main bedrock types.

Figure 5-AD: A composite set of maps locate the areas of detailed sonar surveying in Lake Okanagan. Mosaics compiled from individual sonar records are shown for two locations offshore from Okanagan Mountain Park. The mosaics are approximately 1.5 kilometres long.

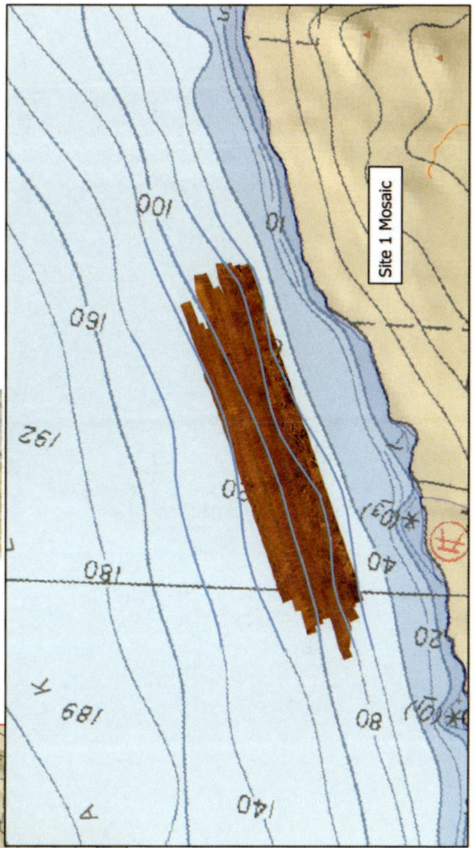

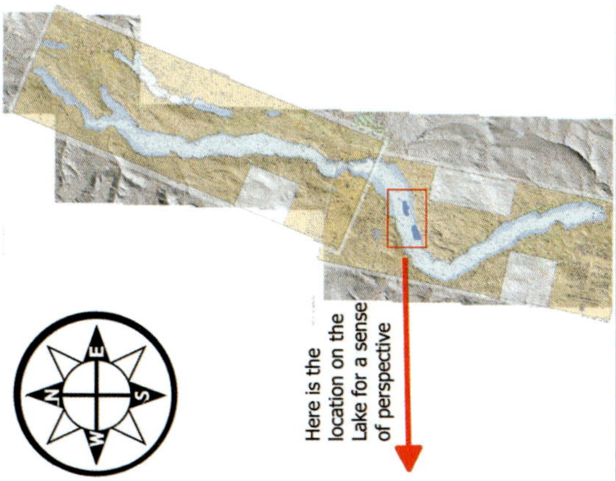

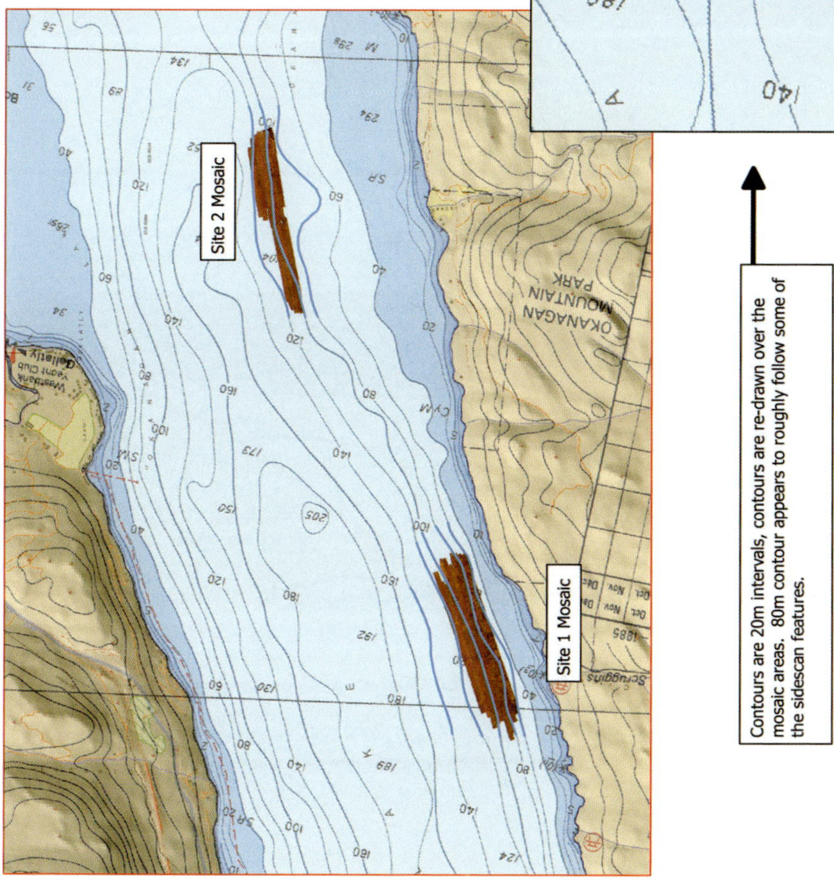

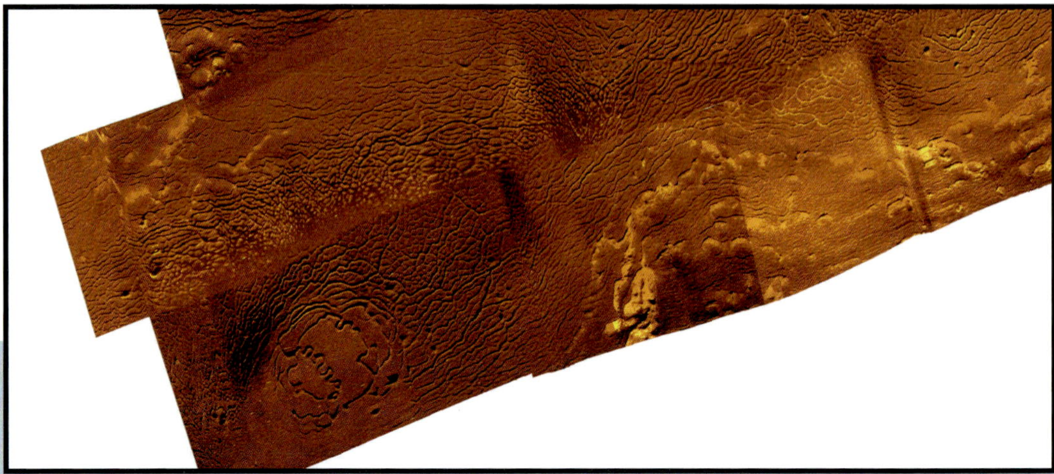

Figure 6-AD: A small part of one of the mosaics is shown here. Distinct panels of the sonor records are noticeable and measure approximate 70 metres long by about 30 metres wide. Three patterns are shown: 1. Bizarre circular furrowed pattern in the lower left ("Ogopogo art"); 2. Extensive furrowed pattern above with large curvatures and pits (dark holes); 3. Bright folded bands in centre-right of the image that may be reflecting bedrock structure or perhaps moraine ridges in shallower water

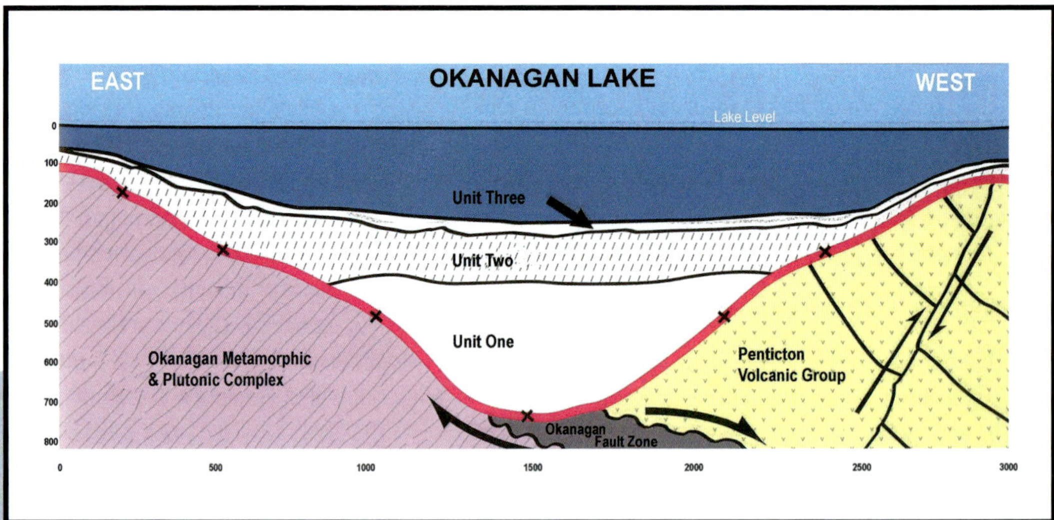

Figure 7-AD: A south-viewing cross-section of Lake Okanagan derived from seismic surveys (see Figure 6-5) shows three major unconsolidated sediment units above the bedrock. This also shows the Okanagan Valley Fault zone with old metamorphic rocks on the East and younger volcanic rocks (Penticton Group) on the West. The sonar images shown are on the surface of Unit Three.

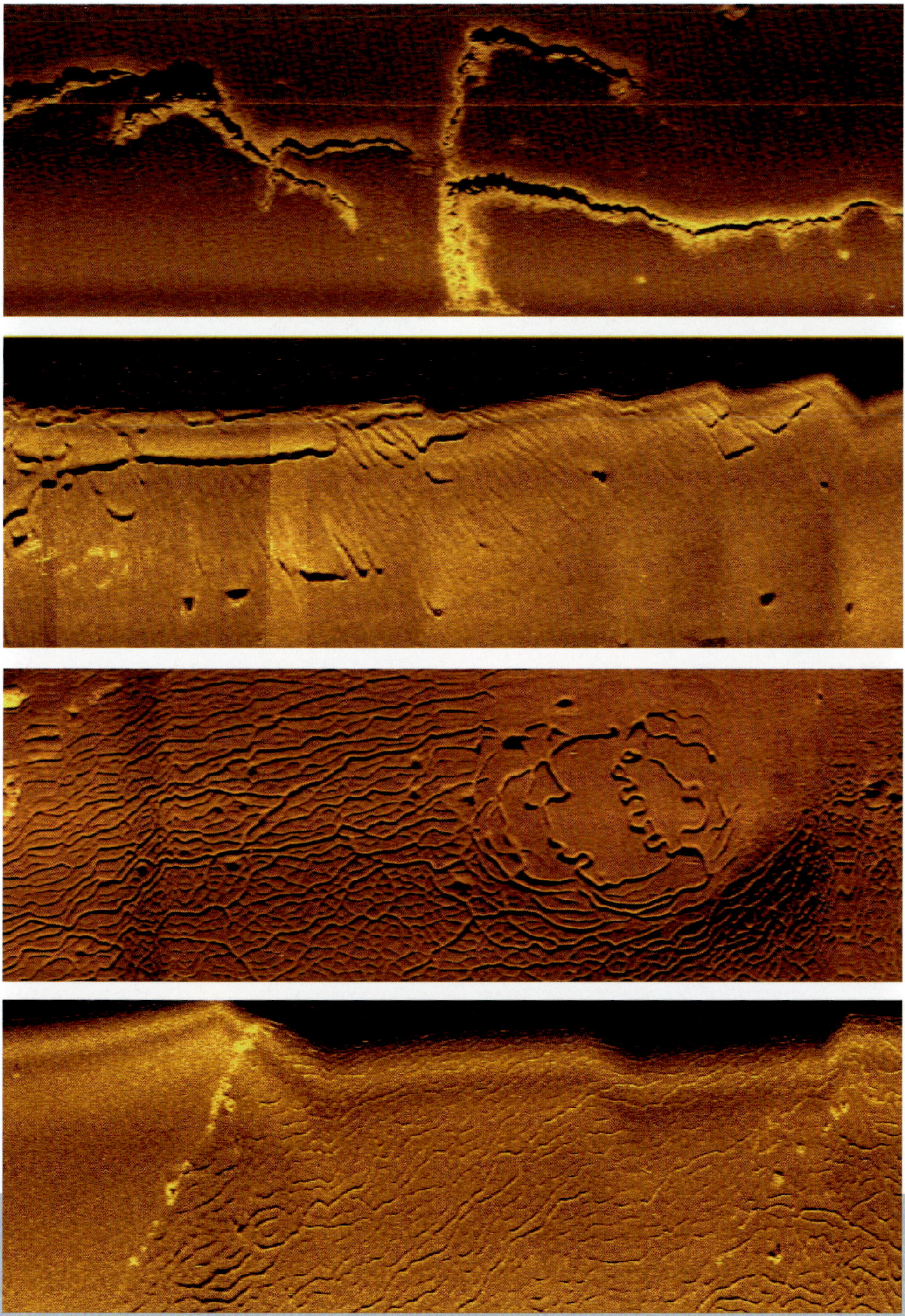

Figure 8-AD: Okanagan Lake Bottom Features: (a) These appear to be sizable fractures a few metres wide and occur in an otherwise slightly reticulated patterned surface. (b) These features appear to be small shallow slides. Ignore the black part of the images; only one side of the sonar tool was active. (c) Here is more detail of the same "Ogopogo art" image shown in Figure 6-AD. The rill patterns are about one half metre wide and pits are a metre wide in places. (d) A line separates a reticulated plain from a near-featureless plain, and is possibly a fault. Images are approximately 20 metres by 50 metres.

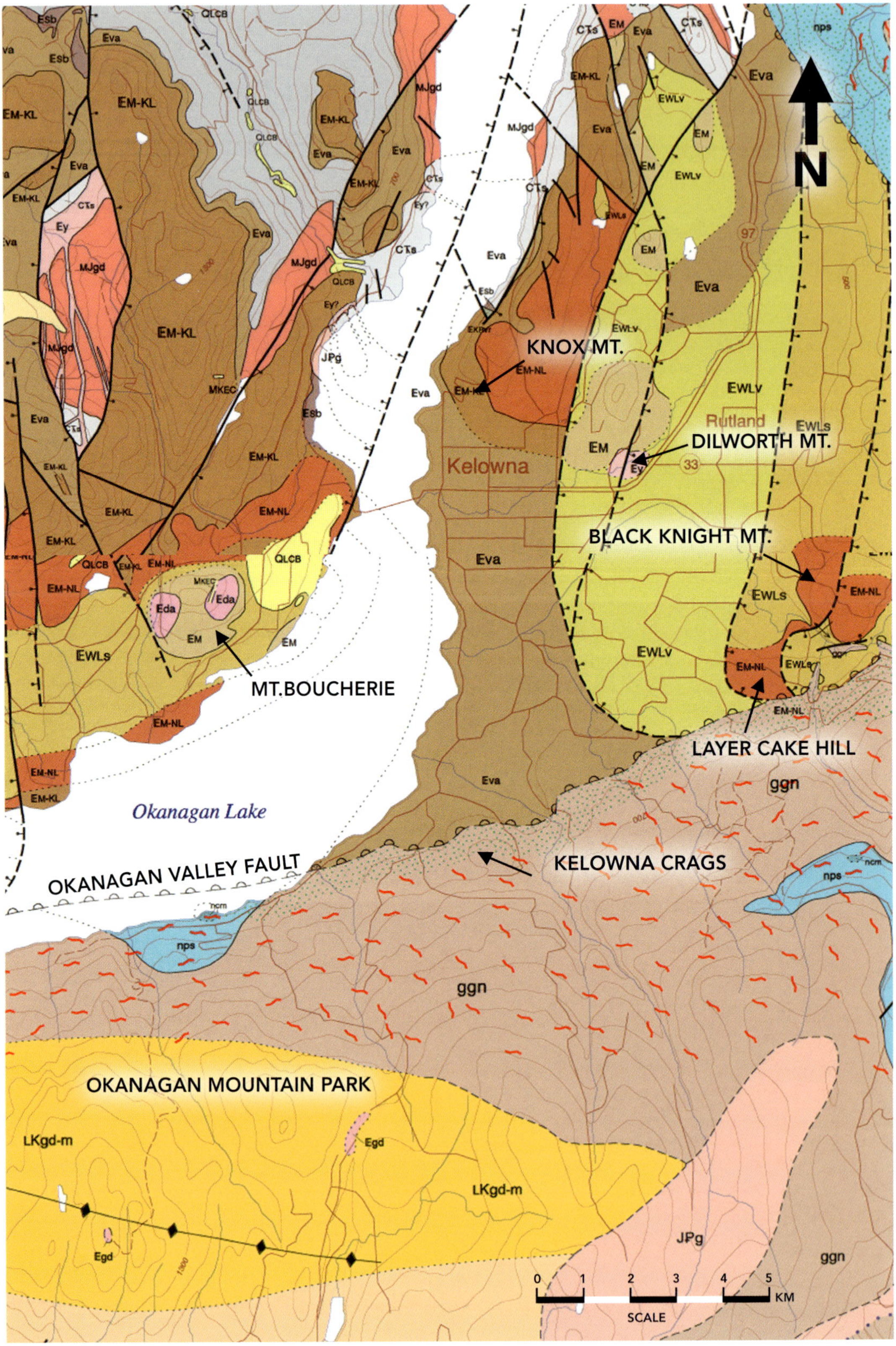

Figure 9-AD: Bedrock map and legend modified after Geological Survey of Canada. Open File 6839

LEGEND

QUATERNARY

QLCB	Lambly Creek Basalt, basalt, massive, average age: 0.97 Ma (Roed et al, 2014). Overlies Westbank First Nation Till.

TERTIARY

MIOCENE AND PLIOCENE

MPK	KALLIS FORMATION, columnar basalt (age: 7.3 - 2.9 Ma).

MIOCENE

MKEC	KING EDWARD CREEK FORMATION, basalt, lava and dykes (age: 20.4-10.4 Ma).

EOCENE

Ey	CORYELL SYENITE, syenite, pink to buff, quartz monzanite, dykes; feeder pipes to Marron Formation (age: 53-47 Ma).
Eda	Dacite, black, columnar-jointed; pipes on Mt. Boucherie. ANGULAR UNCONFORMITY
EWLv	WHITE LAKE FORMATION, mainly breccia, volcanic flows and pyroclastic deposits, some conglomerate, sandstone, river sediments during waning vulcanism.
EWLs	WHITE LAKE FORMATION, mainly conglomerate, sandstone, siltstone, minor lignite, fluvial and lacustrine deposits, plant fossil remains; volcanic interbeds at White Lake (type section).
EM	MARAMA FORMATION, Black Knight Mountain, Layer Cake Hill, Mount Boucherie, Giants Head, dacite domes, pipes, columnar jointing, andesite and breccia (age: 48.4-45.2 Ma).
EM-NL	Nimpit Lake Member, lava, trachyte, minor breccia.
Eva	Lava, andesite, brown, quartz filled amygdules, breccia.

LATE CRETACEOUS

LKgd-m/LKgd-mb	Granodiorite, hornblende, foliated (age: 69-67 Ma)

JURASSIC TO TERTIARY

JPg	OKANAGAN BATHOLITH, granodiorite, granite, medium to coarse grained, unfoliated mainly.

LATE JURASSIC

LJpqm	Quartz monzanite and granite.

MIDDLE JURASSIC

MJgd	Granodiorite, quartz diorite,, granite, moderately foliated.

EARLY JURASSIC

EJgd	BROMLEY/PENASK BATHOLITHS, granodiorite, quartz gabbro, garnet skarn.

UPPER TRIASSIC OR OLDER

CTs	Slate, black, phyllite, argillite, limestone, greenstone, fossils.

CARBONIFEROUS TO MIDDLE TRIASSIC

CTHRs	HARPER RANCH GROUP, siltstone, sandstone, conglomerate, tuff, andesite, minor marble, skarn.

PALEOZOIC (may include PROTEROZOIC)

nps	Schist, paragneiss, interfoliated, garnetiferous, minor quartzite and marble.
ggn	OKANAGAN GNEISS (MONASHEE GNEISS), orthogneiss, granite, mylonitic, schist, interfoliated with Okanagan Batholith, minor gabbro, amphibolite and pegmatite.

ROCK TYPES OF THE OKANAGAN

The Kelowna area is dominated by three vastly different rock assemblages. From oldest to youngest these are: (1) Okanagan Gneiss (Monashee Gneiss, ggn), possibly ranging from Precambrian to Devonian in age, that occurs in Okanagan Mountain Park and parts of the Okanagan Highland east of Okanagan Lake. These metamorphic rocks have been intruded by Jurassic to Late Cretaceous aged granodiorite and granite plutons and batholiths (LKgd-m and JPg). The entire group has been referred to as the Okanagan Metamorphic and Plutonic Complex (Okultich, 1984) and is in fault contact with all other younger rocks along the Okanagan Valley Fault. (2) The Harper Ranch Group occurs in the Bear Creek valley with an east-west band also exposed at Vernon. It ranges from Carboniferous to Triassic in age and is composed of a diverse assemblage of siltstone, sandstone, conglomerate, some marble and minor volcanic rocks. One interpretation of this unit is that it represents a marine offshore basin along the pre-Mesozoic edge of the North American plate at least 245 million years ago. (3) The most extensive bedrock in the Kelowna area consists of numerous volcanic deposits formed 45 to 60 million years ago in the Eocene Epoch of the Tertiary Period. They are referred to as the Penticton Group and occur on both sides of the valley. Many record explosive activity (ash) from what must have been towering volcanoes with thick and extensive lava flows. Some examples locally are Knox Mountain, Black Knight Mountain, Dilworth Mountain and Boucherie Mountain. As volcanism waned a huge river developed and carved a deep

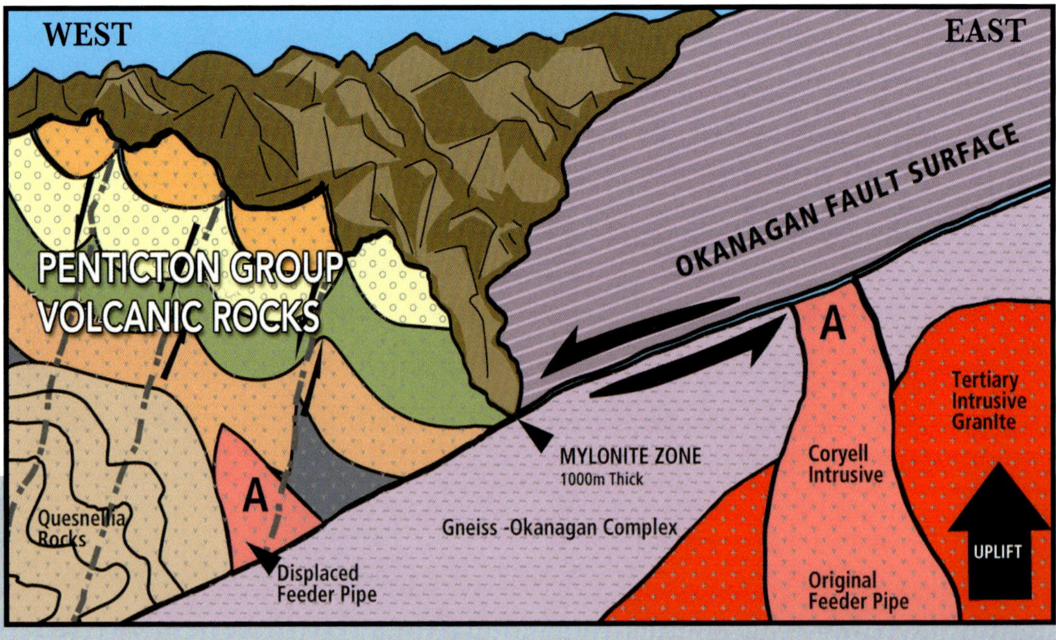

Figure 10-AD: A graphic image showing the main elements of the Okanagan Valley Fault. View is toward the North. The current concept is that the fault occurred before, during and possibly also after the Eocene volcanic interval, and the volcanics slid off to the West.

Figure 11-AD: Basalt of the Chilcotin Volcanics displays columnar structure on the West Kettle Forest Service Road just off the Big White road along the West Kettle River.

canyon that filled with carbonaceous siltstone, sandstone, conglomerate and even volcanic debris represented today by the White Lake Formation (EWLs). Thin lignite coal seams and fossil plant fragments also occur.

OKANAGAN VALLEY FAULT

All of the Eocene volcanic activity is related to the initiation of the Okanagan Valley Fault, thinning of the crust to tap magma chambers, and uplift of the Okanagan Metamorphic and Plutonic Complex. The latest concept is illustrated in Figure 10-AD. As the metamorphic rocks rose from a depth of perhaps 20 kilometres, volcanic rocks on top of it slid off towards the west. The feeder roots of some of the volcanoes have been found in the metamorphic terrain, so this is one of the reasons this theory has been put forward.

RISE OF THE ROCKY MOUNTAINS

All of the foregoing bedrock has been affected by folding and faulting beginning in the late Eocene (Columbian Orogeny) and culminating later in the Oligocene 23 to 37 million years ago with the gradual rise of the Rocky Mountains.

CHILCOTIN VOLCANIC ROCKS

The Interior of British Columbia was substantially eroded during the Miocene and Pliocene and was regionally affected by a much different interval of volcanic activity. At a time when relief was on the order of 300 metres or so the lowlands were "flooded" with highly fluid basalt escaping from fractures and pipes. This period is recorded by the Chilcotin Volcanics Rocks that range in age from 15 million to less than one million years old. Many of these volcanics can be found in the uplands of the region (Figure 11-AD). At West Kelowna the youngest of these is the Lambly (Bear) Creek Basalt (QLCB), a lava flow which erupted near the headwaters of Bear Creek at least 762,000 years ago and flowed 24 kilometres to the base of Mount Boucherie. It now caps the terrace beneath Lakeview Heights on the west side of Okanagan Lake and overlies the Westbank First Nation Till.

Surficial Geology Map of the Kelowna Area

A coloured surficial geology map of the Kelowna area appeared in 2007 (Paradis, 2007). It covers 1848 square kilometres stretching from the latitude of Peachland north to the middle of Kalamalka Lake. Surficial deposits refer to all unconsolidated material and include glacial deposits such as till, sand and gravel from meltwater outwash, glaciolacustrine silt and clay, alluvial deposits in streams and terraces, organic deposits in muskegs and swamps, and windblown sand. Bare rock is also shown on surficial geology maps but only rarely is it lithologically identified as granite, sandstone or whatever. Refer to the bedrock map of Okultch (2013) for lithology. Also, the map in Figure 5.5 is still a good summary of surficial deposits.

A considerable amount of detail of the surficial geology on Paradis' map is shown for the Kelowna area. Of particular interest are scarps delineating terraces and slopes that help to identify erosional levels, also the sites of potential landslides. Glacial features include ice scoured grooves and drumlins that record the southward movement of glaciers, mainly the Fraser Glacier, the most recent. Much of the upland terrain is identified having a blanket or thin veneer of till overlying bedrock. Numerous meltwater channels are shown providing some insight into how the last glacier melted. Of particular interest are the of very large glaciofluvial sand and gravel deposits that identify major entry points into Glacial Lake Penticton. From south to north these include Trepanier, Belleview, Powers, MacDonald and Bear Creeks, a huge deposit coming out of the Ellison and Woodlake valley beside the airport, and another one along the lake at Shorts Creek associated with the alluvial fan of Fintry delta.

The flatland of Kelowna is shown as an alluvial plain representing a complex of meandering alluvial channels, oxbow and organic deposits formed by Kelowna and Mission Creeks

and other smaller creeks. The flat nature of this landform inherently implies a sustained erosional period of flood waters that "levelled" the land when Glacial Lake Penticton finally drained about 10,000 years ago and established the Okanagan River drainage system. Over 1000 metres of unconsolidated sediment, from to older glaciations and interglacial intervals, may lie below the flatland of Kelowna.

Kelowna Crags

A description of Kelowna Crags, located in Cedar Mountain Regional Park off Chute Lake Road, was suppose to be in the original book so is added here. Recently the Johns family has donated 800 acres to the park and it is named after this significant donor. It is known as the Johns Family Nature Conservancy Regional Park.

The Crags are perhaps the most popular rock-climbing locality in the central Okanagan. There are stacks, chutes, chimneys, and shear cliffs, made up of granitic gneiss that provide many challenging climbs in the area. The "Crags" are a series of steep rock ridges, located in the Okanagan Highland, that trend in an East-West direction (Figure 12-AD. and are composed of the Monashee Gneiss.

The rocks, now assigned to the Okanagan Metamorphic and Plutonic Complex (Okulitch, 2013), are thought to have formed as a result of subduction during Devonian time (400 million years ago). They remained deeply buried to a depth of 20 to 40 kilometres and began to rise along the west-dipping Okanagan Valley Fault (Figure 10-AD), before, during or after eruption of the volcanic rocks that make up the 45 to 60 million year-old Eocene volcanic terrain of the Okanagan. The rock ridges of the Monashee Gneiss have been subjected to weathering of course but were deeply gouged out by a number of glacial episodes and eroded by meltwater streams as the glaciers wasted away. The result has been the creation of numerous cliffs and high pillars of rock separated by gently sloping valleys. Most of the rock ridges have low northern slopes and very steep southern slopes, vertical in places, likely due to the plucking action of at least three separate glaciations that flowed south.

Figure 12-AD: An acrylic painting by M.A.Roed showing the type of terrain at the Kelowna Crags, a popular climbing locality just off Chute Lake Road in the newly formed Johns Family Nature Conservancy Regional Park (use to be Cedar Mountain Regional Park). Black Knight Mountain is in the distance.

William R. Bennett Bridge

by Uthaya M. Uthayakamur, P.Eng.

The new five-lane William R. Bennett Bridge (Figure 13-AD) across Okanagan Lake was opened to traffic in May 2008. This bridge links the City of Kelowna on the east to West Kelowna on the west. The previous crossing, a three-lane Okanagan Lake Bridge, was constructed in the 1950s and was in service for over 50 years. The old bridge was located to the immediate south of the new bridge and included a lift span near its eastern end to allow boat traffic beneath the bridge. The operation of the lift span required closing the bridge, which tied-up traffic within the City of Kelowna. In addition, the capacity of the three-lane bridge was inadequate for modern traffic demands, and required significant upgrades to meet new seismic design standards (potential earthquakes). Further, the cost of maintenance and repair of the aging structure was consistently increasing.

Considering the above, the British Columbia Ministry of Transportation and Infrastructure decided to replace the aging bridge with a new one. The design and construction commenced in 2005 and was completed in three years. After opening in 2008, the old bridge was demolished.

Figure 13-AD: This is a fantastic northeasterly view of the William R. Bennett Bridge provided by Mike Bidden, a Penticton artist, author and Search and Rescue Officer who flies his own plane on rescue and photographic ventures. The five piers constructed to support the navigation route beneath the bridge are plainly visible (see the following photos). What a great looking modern structure with downtown Kelowna and many of the gorgeous landforms of the valley in the background.

Figure 14-AD: A photo of the facilities in place to drive the large diameter piles on the west side to support the piers for the elevated fixed portion of the bridge that allows navigation of boat traffic below the bridge.

LAKE SETTING AND DESIGN CONSIDERATIONS

Okanagan Lake occupies a relatively narrow fjord-like trench. The lake is shallow at the location of the bridge crossing, approximately 45 m deep near the mid-length of the bridge. Water depths up to 240 m have been reported elsewhere. Although bedrock is exposed on the western bank of the lake just south of the bridge location, the depth to bedrock or very dense sediments beneath the lake crossing is estimated to be more than 200 m. The subsurface sediments along the bridge alignment consist of very soft glaciolacustrine or lacustrine deposits. In the vicinity of the eastern side of the bridge the lacustrine sediments are inter-layered and overlain by deltaic silty sand and sand layers deposited by Kelowna Creek.

The presence of deep water and very thick compressible and soft sediments made it impractical to construct a fixed bridge rather than a floating bridge. A fixed bridge would have required large pylons or piers to support a large span bridge across the lake. Construction of pile foundations for such large pylons would have been uneconomical. Therefore, a design which included an elevated-fixed structure supported on pile foundations over shallow water and a floating structure supported on pontoons over deep water was chosen. The elevated structure allowed provision for a navigation channel.

BRIDGE CHARACTERISTICS

The new five-lane bridge crossing consists of three westbound lanes but only two eastbound lanes to Kelowna (Figure 13-AD) because current road infrastructure through the City of Kelowna cannot handle more than two inbound (eastbound) lanes.

The total length of the new bridge between the west and the east abutments is 1060 metres. Key components of this new crossing include: (1) a 190 metres long west approach embankment; (2) the west abutment and five

Figure 15-AD: Here are the finished foundation structures, the pile caps, that will support the elevated-fixed portion of the bridge. The old bridge west approach embankment can be seen to the right part of the photo.

Figure 16-AD: This shows the initial construction of a pontoon section and reinforcement complex of the basal slab.

Figure 17-AD: An example of a huge concrete anchor that was employed for the previous floating bridge. Other original anchors are still employed for the present bridge to anchor the 750 metre floating section of the bridge. This anchor is on display at the corner of Ellis Street and Harvey Avenue (Highway 97) in downtown Kelowna and has been for 20 years or so. See also Figure 1-11.

piers over a length of 275 metres; (3) the west transition span of 52 metres; (4) a 730 metres long floating section supported on nine pontoons; (5) a 50 metres long east transition span; (6) the east abutment; and, (7) a 300 metres long east approach embankment.

The west abutment, the five piers and the east abutment are supported on 610 mm and 914 mm diameter steel pipe piles driven to depths of 30 to 50 metres below the lake bottom (Figures 14-AD and 15-AD). The piles were driven and then at least two piles at each pier and at each abutment were tested to confirm the design calculations on the capacity of piles. The design was such that the maximum load carried by each pile was limited to half of the measured capacity.

The 730 metres long floating section consists of nine reinforced concrete pontoons 25 m wide, about 5 metres deep and 30 metres to 90 metres in length (Figures 16-AD, 18-AD). The pontoons are interconnected along the length of the bridge and then fixed to the ends of the bridge. Further, high-strength steel cables and concrete anchors (Figure 17-AD) embedded within the lake bottom are utilized to secure the pontoons and maintain the horizontal alignment of the bridge. The pontoons can move up or down with the seasonal water level in the lake.

PONTOON CONSTRUCTION AND PLACEMENT

The pontoons were constructed at a custom built temporary dry dock near Bear Creek Provincial Park, on the west side of the lake (Figures 16-AD, 18-AD). The dry dock was approximately 100 metres long, 30 metres wide and 7 metres deep. Following construction of a pontoon, the dry dock was flooded by opening a gate which allowed water to enter and float the pontoons. The pontoons were then pulled by boat out to the bridge location and secured to anchors. The dry dock was decommissioned and filled with sand after completion of the project.

SITE PREPARATIONS FOR BRIDGE APPROACHES

The presence of very soft and compressible sediments underlying the bridge approach embankment required "preload treatment". Preload treatment is a process in which the highway embankment is built several months to years ahead of schedule and is allowed to undergo settling before completion of construction and paving. A "surcharge", consisting of additional fill above the design road elevation, is used to accelerate the settlement during the preload treatment phase. Basically what this does is squeeze water out of the subsurface sediments which imparts more strength to them.

Highway embankments are commonly constructed using mineral earth fill, typically consisting of sand and gravel. However, the lake bottom sediments are highly compressible, weak and not suitable or capable of supporting high embankments constructed entirely using mineral earth fill. Therefore, options of embankments constructed using light weight fill material were considered. Light weight fill materials such as cellular concrete, pumice, hog fuel (wood waste) and Expanded PolyStyrene (EPS, commonly known as styrofoam blocks) have been used for highway embankments throughout BC and in many parts of the world. EPS was chosen as the light-weight fill for this project following analysis of the options.

Figure 18-AD: An incredible photo of a pontoon construction phase showing floatation chambers just prior to concrete mantling. Two completed pontoons are anchored outside the dry dock ready to be floated to the bridge site for emplacement.

Figure 19-AD: Here are the slabs of Styrofoam (EPS) used to lighten the embankment load on the west approach to the bridge.

Figure 20-AD: The construction of the EPS (styrofoam) embankment and the bridge girders to fit eventually onto the five piers for the elevated navigation facility are shown in this photo. Two new pontoons are anchored closeby for later installation.

The density of the EPS used in highway embankments is in the order of 20 to 23 kg/m3, about 1/100th of the density (lighter than) of mineral earth fill and 1/50th of the density of water. As the density of EPS is significantly less than that of water, floatation potential of the embankment for the maximum possible water level had to be considered in the design.

DETAILS OF THE DESIGN FOR APPROACHES

Following detailed settlement, slope stability and seismic deformation analyses, the west approach embankment was designed and constructed using a combination of mineral earth fill and EPS fill. The EPS embankment was approximately 130 m in length and 25 m in width (Figure 20-AD). The thickness of the EPS fill varied from 0.6 m to 6 m over the length of the embankment. The EPS fill was supplied in approximately 1.5 m x 1.5 m x 4.5 m sized blocks cut as required at the site to fit the embankment geometry (Figure 19-AD). After placement, the EPS fill was wrapped with 10 millimetres thick polyethylene sheets and a minimum of one metre of granular fill was placed over the polyethylene cover. The purpose of the polyethylene cover is to protect the EPS fill from potential gasoline spill or leakage as contact with gasoline can damage the EPS. Construction of the EPS embankment and bridge girders is shown in Figure 20-AD, and two new pontoons are shown anchored for installation later.

The east approach embankment was constructed entirely using mineral earth fill (i.e.: sand and gravel) following preload treatment. Light-weight fill was not required for the east approach embankment because the subsurface sediments of Kelowna Creek deposition are not as compressible as those on the west side. Construction of the embankments for preload was started in June 2005 and was completed in October 2005. Preload treatment was completed in December, 2006, approximately 500 days after the start of preload construction. Settlement (dewatering progress) of the preload fill and the subsurface soil layers were monitored using shallow and deep settlement gauges at weekly to bi-weekly intervals during the treatment phase. In addition, horizontal deformation and changes in the water pressure within the subsurface soils were monitored using appropriate instrumentation. After the preload treatment and settlement, the remaining surcharge fill was removed and construction of the final road embankment and paving were completed.

On-going post-construction settlement monitoring program shows that the measured settlement magnitudes and rates of settlement are within the predictions made during the design phase, indicating satisfactory performance.

References for Addendum

Okulitch, A.V. (Comp.), 2013. Geology, Okanagan Watershed, British Columbia, Geological Survey of Canada, Open File 6839, scale 1:I00 000. Doi:10.4095/292220. Order on-line at Geoscan.

Paradis, S.J., 2007. Surficial Geology, Kelowna, British Columbia. Geological Survey of Canada, Open File 6146, scale: 1:50,000. Order on-line at Geoscan.

Roed, Murray A., 1995.Geology of the Kelowna Area and Origin of the Okanagan Valley, British Columbia. Kelowna Geology Committee, 183p.

Roed, Murray A., Fulton, Robert J., Editors, 2011 Okanagan Geology South, Geologic Highlights of the South Okanagan, British Columbia, Okanagan Geology Committee, 238p.

Roed, Murray A., Rene W. Barendregt, Jeff A. Benowitz, C.A.S. Smith, P.T. Sanborn, John D. Greenough, Crystal Huscroft, Paul W. Layer, Rolf W. Mathewes, and Dez Tessler, 2014. Evidence for an Early Pleistocene glaciation in the Okanagan Valley, southern British Columbia, Canadian Journal of Earth Sciences, 51: 125-141.

(Uthaya) M. Uthayakamur and Ernest Naesgaard, 2011. Geotechnical Design for the William R. Bennett Bridge, Kelowna, British Columbia. Proceedings of the Pan-American and Canadian Geotechnical Society Conference, October 2011, Toronto, ON.

A view across Okanagan Lake from Tantalus Vineyards on DeHart Road. The view has eliminated all other changes to the land and features Boucherie Mountain across the lake and the low terrace of Lakeview Heights underlain by the Lambly Creek Basalt valley lava low which overlies the till of the West Kelowna Glacier Advance that occurred over one million years ago. Acrylic painting by M. Roed.

About the Authors

James Baker, B.A., M.A. James is a professor Emeritus in anthropology at Okanagan University College where he taught archaeology and anthropology for 26 years. James' speciality is the archaeology and ethnography of the Plateau Culture Area. Presently, James is Mayor of Lake Country.

Dave Capstick, B.Sc., deceased. Dave was originally a petroleum geologist. Re-locating to Kelowna in 1971, he was employed at Big White, and spent 10 years in plant nursery operations. He was an avid outdoor enthusiast and lapidary specialist.

Don A. Dobson, B.Ap.Sc., P.Eng. Don is an engineering consultant with a specialty in hydrology. He has worked for government, community water districts and the forest industry. Don and his wife Terry and family live in Kelowna.

George Ewonus, B.A., M.Ed.(Med.Ed.), is the founding Director of the College Board in Canada and makes his home in Victoria with his wife Lindy and family. George has a keen interest in archeology, has managed digs in the Okanagan Mountain Provincial Park and has prepared papers on his findings.

Paul Ewonus, B.Sc., is studying for a Master's degree in archeology at McMaster University in Hamilton. Paul grew up in the Okanagan and has consulted in archeology since 1998. He lives in Dundas, Ontario.

John D. Greenough, Ph.D., P.Geol., FGAC, is a founding professor in the Department of Earth and Environmental Sciences at UBCO. He has published on the geochemistry of everything from rocks to priceless artifacts, and Canadian wines. He lives and works in Kelowna.

Brian Hughes, B.Sc., B.Ed., P.Geol., is a geological consultant for mineral and petroleum exploration and deep basin horizontal drilling. Brian grew up in Penticton and has resided in Kelowna with his wife Nicole and family since 1985.

H.A. (Herb) Luttmerding, B.S.A., P.Ag., deceased. Herb was an agricultural soils and terrain specialist with twenty-five years of government service. He has published numerous soils reports and maps in British Columbia. His wife Mary lives in Kelowna.

Leanne Mallory-Greenough, Ph.D., is an archaeologist and geochemist. Her research focuses on the use of geochemistry to study the origin of everything from artifacts to the terrestrial planets. She lives in Vernon and teaches at Kalamalka College.